SpringerBriefs in Physics

For further volumes:
http://www.springer.com/series/8902

Andrea Macchi

A Superintense
Laser–Plasma Interaction
Theory Primer

 Springer

Andrea Macchi
National Research Council
National Institute of Optics (CNR/INO)
Pisa
Italy

and

Department of Physics "Enrico Fermi"
University of Pisa
Pisa
Italy

ISSN 2191-5423   ISSN 2191-5431   (electronic)
ISBN 978-94-007-6124-7   ISBN 978-94-007-6125-4   (eBook)
DOI 10.1007/978-94-007-6125-4
Springer Dordrecht Heidelberg New York London

Library of Congress Control Number: 2012955408

Printed on acid-free paper

Springer is part of Springer Science+Business Media (www.springer.com)

Preface

In the past decade, the field of superintense laser-plasma interactions has been exceptionally vital. Progress in wakefield acceleration of electrons has conquered the cover of *Nature*. The observation of high brilliance beams of multi-MeV protons from solid targets has stimulated an enormous amount of work aiming at applications in technology and medicine. Plasma-based schemes have been proposed to compress in time and space laser pulses achieving extremely high field strength, approaching the Schwinger limit at which the vacuum becomes unstable for pair production. On these playgrounds, laser-plasma scientists have been more and more in interaction with colleagues from other communities, and topics such as radiation friction and quantum electrodynamics effects have become of central interest. This research effort has been, and will be further sustained by developments in laser technology. There is at present an impressive number of either planned or running projects based on high power laser installations, including both large international facilities which will provide the highest laser intensities ever produced, and smaller scale systems bringing petawatt power on a national scale.

As several new research perspectives are opened, it is expected that many students and young researchers will need training on the basics of laser-plasma interactions, and that some experienced researcher will be willing to have a quick reference to such basics. This book aims to provide a very compact theoretical introduction to the field, at a tutorial level which should be adequate for third-year undergraduate students in physics or engineering. The reader is assumed to know classical electrodynamics and basic relativistic dynamics, while some elementary notions of statistical mechanics and hydrodynamics would be useful but probably not necessary. In particular, the reader may know little or nothing about a *plasma* beyond the rough notion of a piece of ionized matter, where electrons and ions move freely into electromagnetic fields. Concerning the mathematical complexity of the included material, I assumed no more than a basic background in calculus at the level of third-year undergraduates, and tried to present the calculations in a self-consistent manner. In this way, the book could also serve as a short, example-based (and definitely far from rigorous) introduction to some nonlinear methods in theoretical physics.

The major benchmark for this book has been a course of *Relativistic Plasmas* I have been giving in the Master program at the Department of Physics of the University of Pisa since 2009. The title of the course indicates that laser-plasma interaction offers examples of "relativity in action" in many-body systems dominated by collective behavior. In particular, these are the only systems of such type which can be investigated on a laboratory scale, and are of growing interest as test beds for the dynamics of plasmas in remote astrophysical environments.

There are, of course, good books on laser-plasma physics, which are mentioned in the Introduction and where an extended and detailed description of high intensity laser-matter interactions may be found. Almost unavoidably, the more comprehensive a book is, the less it may cover the most recent advances in the field. These latter are typically summarized in review papers, which however are not always at the tutorial level that would be desired by newcomers. The present book might give some essential notions which make the reading of more complete material easier. The need to be very concise has imposed a strong selection of contents to be included, so that there are important topics which are not covered at all, and the number of cited references is very limited. Fortunately, for each major topic there are recent reviews where much more complete bibliographies can be found.

In our selection, we took into account some of the most either recurrent or intriguing questions and doubts which arised, in our experience, while discussing with colleagues from the laser-plasma community. For such discussions, for posing intellectual challenges and for advice, collaboration, encouragement, support, and much more through the years, thanks go to Dieter Bauer, Carlo Benedetti, Marco Borghesi, Sergei V. Bulanov, Federica Cattani, Francesco Ceccherini, Carlo A. Cecchetti, Tiberio Ceccotti, Piero Chessa, Enrique Conejero-Jarque, Fulvio Cornolti, Antonino Di Piazza, Satyabrata Kar, Tatyana V. Liseykina, Pasquale Londrillo, Peter Mulser, Matteo Passoni, Francesco Pegoraro, Kevin Quinn, Caterina Riconda, Lorenzo Romagnani, Hartmut Ruhl, Gianluca Sarri, Andrea Sgattoni. Of course, the greatest stimulus for this book came from the students attending my lectures or working under my supervision at all levels (B.Sc., M.Sc., and Ph.D.) in laser-plasma interactions. Some of them, through their assignments or thesis work have provided material which is included in this book, and have also helped me to find misprints and errors in the draft. Many thanks and best wishes for the future thus go to Alessandra Bigongiari, Marta D'Angelo, Simone De Camillis, Paolo Dell'Osso, Camilla Galloni, Mattia Lupetti, Amritpal Singh Nindrayog, Elia Schneider, Matteo Tamburini, Sara Tuveri, Silvia Veghini.

Special thanks go to my family and my friends, with apologies for the extra time devoted to work while preparing this book. Finally, thanks to Samuele Landucci (1969–1995) whose friendship was so essential during our early years in Physics.

Lucca and Pisa, Tuscany, Italy   Andrea Macchi
October 2012

Contents

Chapter 1
Introduction

The present book deals with laser-plasma interactions. The phenomena which are
described in the book occur when a laser pulse interacts with matter which can be
considered as a *plasma*, i.e. which is ionized during all the interaction, so that the
optical response of the medium is determined by the motion of free electrons. By
superintense laser pulse we will mostly mean that the field amplitude is high enough
that the electrons will be driven at velocities close to the speed of light c, so that
their dynamics must be described by using special relativity. We will briefly refer to
a *laser-plasma* as a plasma interacting with laser light.

The key point in the physical description of the plasmas considered in the present
book is the assumption that the dynamics is dominated by *collective* behavior rather
than by local interactions between neighboring particles. This means that on each
particle the mean electromagnetic (EM) field, whose sources are the average number
and current densities, is much larger than the fluctuations on a small spatial scale.
In other words, the force on each particle due to the fields created by all the charges
in the system plus the external fields is more important than the forces exerted by
the nearest neighboring particles. The latter forces may be treated as *collisions*,
producing a random scattering of the particles, which is assumed to be less relevant
than the coherent motion in the mean field.

Perhaps a clear example of collective behavior which is of direct relevance in
the present context may be found in a paper by Veksler reporting his vision of
future particle accelerators in 1957 (Veksler 1957). Veksler introduced the concept
of "coherent acceleration" as a mechanism in which the accelerating field on each
particle is proportional to the number of particles being accelerated, and automatically
synchronized in time and space with the particles' location. The techniques for laser-
plasma acceleration of electrons and of ions, to be described in Chaps. 4 and 5, may
be considered as a realization of such coherent acceleration paradigm.

A condition necessary to the dominance of collective effects is that all relevant
frequencies must be larger than the frequency of collisions. In a classical plasma, the
collision frequency is proportional to the cross section of Coulomb scattering (also
named Rutherford scattering) and thus decreases with increasing particle energy.

A. Macchi, *A Superintense Laser-Plasma Interaction Theory Primer*,
SpringerBriefs in Physics, DOI: 10.1007/978-94-007-6125-4_1,
© The Author(s) 2013

Hence, the high energy density in the presence of a superintense EM field quenches the rate of collisions, and the plasma can be considered *collisionless*, like an ideal gas where the particles bounce on containing walls but do not interact with each other. We also add that the assumption of fully ionized matter is consistent with the presence of superintense electric fields which are much larger than the binding field inside atoms, at least for outer electrons. It might be worth adding that we will consider exclusively classical plasmas, neglecting quantum effects. This means that, on the low energy side, the mean energy per electron is much higher than the Fermi energy, while on the high energy side we do not consider quantum electrodynamics (QED) effects. The latter might probably appear the most serious limitation in the context of superintense interactions, since the values of EM fields which can be achieved in a laboratory thanks to the developments in laser technology and nonlinear optical manipulation have been increasing so much in recent years that reaching the QED limit might become possible.

The theoretical background over which models of laser-plasma interactions are built, i.e. the Lorentz force and Maxwell's equations for EM fields plus either kinetic or hydrodynamics equations for electrons and ions, has been known from more than one century at present. However, due to the nonlinear character of the basic equations, the solutions are numerous and extremely various and describe many different phenomena, several of which are probably still unknown. Nonlinearity is integral to any collective system as the force on each particle is determined in principle by the motion of all other particles. In some sense, relativistic laser-plasmas are even more nonlinear than other plasmas, because nonlinearity already appears at the single particle level: the oscillation of an electron in an intense, monochromatic EM field contains higher harmonics of the fundamental frequency, and the oscillation amplitude is a nonlinear function of the field, as will be shown in Chap. 2. On a macroscopic scale, a direct consequence of the relativistic equation of motion for electrons is that the relation between the current density and the EM field also becomes nonlinear. This leads to several "relativistic optics" phenomena, of which the main ones are described in Chap. 3. In addition, some proposed schemes for ion acceleration (Sect. 5.7) and for manipulation of superintense laser pulses (Chap. 6) may be considered as applications of the concept of a relativistic "moving mirror", already introduced by Einstein in his original work on special relativity (Einstein 1905).

Although the basic theoretical foundations have been established since ages, the modeling of superintense laser-plasma interaction offers several examples of subjects of historical controversy. These latter include, already at the level of the single particle dynamics, electron acceleration by electromagnetic waves in vacuum, the relativistic ponderomotive force, and radiation friction. At the level of collective dynamics controversy arises even in electrostatic problems such as the formation of collisionless sheaths and the plasma expansion into vacuum, which are relevant to very different regimes such as plasma discharges (Riemann et al. 2005) and ultracold plasmas (Killian et al. 2007). Again, nonlinearity and the related difficulties in the theoretical description are often the sources of debate. Going into the details of such long-standing problems is well beyond the scope of this book, but we try at least to point out where the difficulties are, and to indicate further reading.

The models of laser-plasma phenomena described in this book might be considered of two types, depending on their relation with experiment. The first type includes models originally introduced to support new experimental proposals, such as, e.g., the use of laser wake waves for the acceleration of electrons (Sect. 4.1) or for the extreme amplification of EM fields (Sect. 6.2). The second type includes models introduced to explain novel and sometimes unexpected experimental results. As an example, this was the case for many theory papers following the observation of high-energy protons from solid targets in 2000 (Chap. 5). The starting points of such work are often classic problems of plasma physics, such as sheath formation and collision-less plasma expansion in vacuum for the case of ion acceleration. Most of the times classic models are not sufficient to describe the experimental results satisfactorily so that additional parameters and complexity must be introduced. This approach often leads to a better agreement between theory and experiment, but also generates a risk that the model becomes too specific and representative of a quite particular regime. Thus, in this book we focus on basic reference models, briefly mentioning directions for their improvement.

The present book is just a short primer, aiming to provide a concise and elementary introduction to laser-plasma interaction theory, without reviewing all the work published in the field. The contents of the book are thus limited and there are many important laser-plasma phenomena which have been either briefly mentioned or not included at all such as, e.g., parametric instabilities or the generation of coherent structures and of quasi-steady magnetic fields. A basic description of these latter phenomena as well as a possibly more detailed and rigorous discussion of the topics of the present primer may be found in more extended books (Kruer 1988; Gibbon 2005; Mulser and Bauer 2010) and in review papers, which are cited in the bibliography of each Chapter. Here we just mention a very recent tutorial overview of the field (Gibbon 2012) whose level and purpose are quite similar to the ones of the present book.

As the focus is on theoretical aspects, present and foreseen applications of laser-plasma interactions are only briefly mentioned. We also do not spend many words in presenting future developments and horizons of the research field, which are summarized in several non-specialized or "news and views" articles as well as in most recent reviews. Concerning experimental works, we cite those that, in our opinion, have been the most relevant also with respect to either stimulating theoretical research or confirming the results of the latter, and we do not enter into the details of experimental techniques. Of course, the reference to experiments is also necessary to convince the reader that the theory presented has some relevance to reality.

We have tried to keep the book at a level that should be accessible to students between the second and the third year of an undergraduate program in Physics. Although basic notions of plasma physics may be helpful, a knowledge of classical electrodynamics and special relativity should suffice. More in detail, the reader should already know about basic classical mechanics, Maxwell's equations for the EM fields, propagation of EM waves in vacuum and in matter, equations of relativistic dynamics, and Lorentz's transformations for the EM fields. In dealing with relativity we prefer to use a non-explicitly covariant notation as it is probably most familiar and intuitive for the average reader. Occasionally, concepts of Lagrange-Hamilton mechanics and mathematical techniques such as the Fourier transform or simple integration in the complex plane have been used, but the reader who has not studied yet such topics may simply skip the related paragraphs. The aim of such parts is mainly to stimulate the interest in managing advanced analytical techniques which have broader applications in theoretical physics. Gaussian c.g.s units are used everywhere, since in the author's opinion it would be strange to deal with relativistic electrodynamics using a system where the electric and magnetic fields have different dimensions and the speed of light does not appear in Maxwell's equations. However, when quoting typical numbers and orders of magnitude, we choose practical "hybrid" units such as $W\,cm^{-2}$ for intensities, μm for lengths, and so on.

The level of detail at which the calculations are presented is such that the reader should be stimulated to reproduce the algebra, but at the same time not frightened by the sudden appearance of complex formulas after "a simple calculation shows ..." statements. For any part which should still sound obscure to the reader, the author warmly welcomes questions and comments.

Throughout the book, we searched for a balance between mentioning topics which are of recent and growing interest, and thus may have not yet extensively presented in textbooks, and selecting examples where significant results may be obtained with a reasonable and self-contained analytical effort. In the course on which the book is based, for the final examination the students are encouraged to go deeper in some topic that might have mostly stimulated their curiosity and is not fully described in the lecture notes. In this way, the students put their hands on some problem by trying to reproduce analytical results or also numerical ones, by writing simple codes or using standard computational packages. This approach is usually welcome by the students and, in the author's opinion, represents the best test of the skills acquired during the course, as well as a real "Introduction" to the active field of superintense laser-plasma interactions.

References

Einstein, A.: Ann. Phys. **322**, 891 (1905)
Gibbon, P.: Short Pulse Laser Interaction with Matter. Imperial College Press, London (2005)
Gibbon, P.: Riv. Nuovo Cim. **35**, 607 (2012)
Killian, T., Pattard, T., Pohl, T., Rost, J.: Phys. Rep. **449**, 77 (2007)
Kruer, W.L.: The Physics of Laser Plasma Interactions. Addison–Wesley, New York (1988)
Mulser, P., Bauer, D.: High Power Laser–Matter Interaction. Springer, Berlin, Heidelberg (2010)
Riemann, K.U., Seebacher, J., Tskhakaya, D.D., Kuhn, S.: Plasma Phys. Contr. Fusion **47**, 1949 (2005)
Veksler, V.: At. Energy **2**, 525 (1957)

Chapter 2
From One to Many Electrons

Abstract As it is customary for several books of plasma physics, we begin with a description of single particle dynamics, focusing on the motion of an electron in a plane wave that is most relevant to the physics we are interested in. This part also serves as a quick review of relativistic dynamics and a warm-up to tackle nonlinear equations. We then introduce the concept of the ponderomotive force and finally discuss the issue of radiation friction, an emerging topic in ultra-relativistic laser plasmas. In the second part of the chapter we briefly review the basic kinetic and hydrodynamic modeling of collisionless plasmas and we also give a quick description of numerical modeling, including the pioneering Dawson model, the particle-in-cell method, and the boosted frame techniques to make simulations easier to perform.

2.1 The Single Electron in an Electromagnetic Field

2.1.1 Non-relativistic Motion in a Plane Wave

Before searching for exact and general solutions, it is useful to start with an approximate ("perturbative") calculation which allows us to softly introduce nonlinear and relativistic effects. Since to the lowest order all nonlinear effects are neglected, it is convenient to use the complex representation for the electric and magnetic fields of a plane wave propagating in the x direction[1]

$$\mathbf{E} = \mathbf{E}(x,t) = E_0 \hat{\boldsymbol{\varepsilon}} e^{ikx - i\omega t}, \qquad \mathbf{B} = \mathbf{B}(x,t) = \hat{\mathbf{x}} \times \mathbf{E}, \qquad (2.1)$$

---

[1] Unit vectors are indicated with a "hat" symbol, so that e.g. a generic position vector $\mathbf{r} = x\hat{\mathbf{x}} + y\hat{\mathbf{y}} + z\hat{\mathbf{z}}$.

A. Macchi, *A Superintense Laser-Plasma Interaction Theory Primer*,
SpringerBriefs in Physics, DOI: 10.1007/978-94-007-6125-4_2,
© The Author(s) 2013

where it is implicit that the physical fields are the real parts of the above expressions. Here, $k = \omega/c$ and $\hat{\boldsymbol{\varepsilon}}$ is the complex polarization vector. For linear polarization along y (z), $\hat{\boldsymbol{\varepsilon}} = \hat{\mathbf{y}}$ $(\hat{\mathbf{z}})$ while for circular polarization $\hat{\boldsymbol{\varepsilon}} = (\hat{\mathbf{y}} \pm i\hat{\mathbf{z}})/\sqrt{2}$ corresponding to counterclockwise and clockwise directions, respectively.

The non-relativistic equations of motion for an electron are

$$m_e \frac{d\mathbf{v}}{dt} = -e \left[\mathbf{E}(\mathbf{r}, t) + \frac{\mathbf{v}}{c} \times \mathbf{B}(\mathbf{r}, t) \right], \qquad \frac{d\mathbf{r}}{dt} = \mathbf{v}, \qquad (2.2)$$

where in general the EM fields depend on the particle position $\mathbf{r} = \mathbf{r}(t)$. To lowest order, i.e. in the linear approximation we neglect the $\mathbf{v} \times \mathbf{B}$ because, for weak fields, $|\mathbf{v}| \ll c$ and also because we are interested in the oscillating motion at the same frequency of the EM wave: if $\mathbf{v}(t)$ is a function oscillating with frequency ω, the $\mathbf{v} \times \mathbf{B}$ product contains both a 2ω oscillating term and a constant (0ω) term, but not any term with frequency ω. The linear solution is thus obtained immediately as

$$\mathbf{v} = -\frac{ie}{m_e \omega} \mathbf{E}, \qquad \mathbf{r} = \frac{e}{m_e \omega^2} \mathbf{E}, \qquad (2.3)$$

where the electric field is taken at the constant x-position of the electron, since there is no force in the longitudinal $(\hat{\mathbf{x}})$ direction. The trajectory is a line for linear polarization, and a circle for circular polarization, as may sound obvious.

Having obtained this simple solution, we can check *a posteriori* that $|\mathbf{v}| \ll c$. Since the peak amplitude is $v_0 = eE_0/m_e\omega$, the linear solution is adequate as long as $a_0 \ll 1$, where the dimensionless parameter a_0 is defined as

$$a_0 \equiv \frac{eE_0}{m_e \omega c}. \qquad (2.4)$$

Now suppose that a_0 is not so small with respect to unity and thus we need to estimate the effects of the magnetic force. The idea is to use an iterative approach with a_0 as the expansion parameter. We assume that to the lowest order in a_0, the electron velocity is given by the linear solution (2.3), and we use this solution to evaluate the next order term. Thus we write for the velocity $\mathbf{v} = \mathbf{v}^{(1)} + \mathbf{v}^{(2)}$ where $\mathbf{v}^{(1)}$ and $\mathbf{v}^{(2)}$ are of order $\sim a_0$ and $\sim a_0^2$, respectively. Inserting this expression into the equation of motion

$$m_e \frac{d(\mathbf{v}^{(1)} + \mathbf{v}^{(2)})}{dt} = -e \left[\mathbf{E}(\mathbf{r}, t) + \frac{\mathbf{v}^{(1)} + \mathbf{v}^{(2)}}{c} \times \mathbf{B}(\mathbf{r}, t) \right], \qquad (2.5)$$

and equating first only the $\sim a_0$ terms and then the $\sim a_0^2$ ones we obtain

$$m_e \frac{d\mathbf{v}^{(1)}}{dt} = -e\mathbf{E}, \qquad m_e \frac{d\mathbf{v}^{(2)}}{dt} = -e\frac{\mathbf{v}^{(1)}}{c} \times \mathbf{B}. \qquad (2.6)$$

From the first equation we simply get the confirmation that $\mathbf{v}^{(1)}$ is the linear solution. Now, assuming *linear* polarization along y ($\hat{\mathbf{e}} = \hat{\mathbf{y}}$), we may write

$$\mathbf{v}^{(1)} = \frac{eE_0}{m_e\omega}\hat{\mathbf{y}}\sin\omega t = a_0 c\hat{\mathbf{y}}\sin\omega t, \qquad y^{(1)} = -a_0\frac{c}{\omega}\cos\omega t, \qquad (2.7)$$

where we took $x = 0$ as the initial position of the electron and switched back to the real part of the fields, to avoid complications with nonlinear terms. Since $\mathbf{B}(x = 0, t) = E_0\hat{\mathbf{z}}\cos\omega t$, the equation for $\mathbf{v}^{(2)}$ becomes

$$\frac{d\mathbf{v}^{(2)}}{dt} = -\hat{\mathbf{x}}\frac{e}{m_e c}(a_0 c\sin\omega t)(E_0\cos\omega t) = -\hat{\mathbf{x}}\frac{a_0^2}{2}c\omega\sin 2\omega t, \qquad (2.8)$$

so that the electron oscillates along the x direction with frequency 2ω:

$$v_x^{(2)}(t) = \frac{a_0^2 c}{4}\cos 2\omega t, \qquad x^{(2)}(t) = -\frac{a_0^2 c}{8\omega}\sin 2\omega t. \qquad (2.9)$$

By eliminating t from the solutions for $y \simeq y^{(1)}$ and $x \simeq x^{(2)}$, and then defining $X = (\omega x/c)/a_0^2$ and $Y = (\omega y/c)/a_0$ it is easy to obtain the trajectory in the following form:

$$16X^2 = Y^2(1 - Y^2), \qquad (2.10)$$

which describes a "figure of eight" as shown in Fig. 2.1. The eight is "thin" since the x/y aspect ratio $a_0/8$ is rather small for $a_0 \ll 1$ as required by our approximation.

For *circular* polarization, with some simple vector algebra we find

$$\mathbf{v}^{(1)} \times \mathbf{B} \propto (-\hat{\mathbf{y}}\sin\omega t \pm \hat{\mathbf{z}}\cos\omega t) \times (\hat{\mathbf{z}}\cos\omega t \mp \hat{\mathbf{y}}\sin\omega t)$$
$$= \hat{\mathbf{x}}(-\sin\omega t\cos\omega t + \cos\omega t\sin\omega t) = 0, \qquad (2.11)$$

so that the trajectory is unaffected by magnetic force effects with respect to the first order of approximation: the electron still performs a circular trajectory of radius $a_0 c/(\sqrt{2}\omega)$.

Fig. 2.1 The "universal" figure-of-eight *curve* (2.10) describing the trajectory of an electron in a monochromatic plane wave, in the reference frame where there is no average drift along the propagation direction (x)

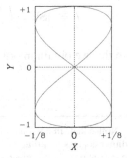

In principle we do not need to stop our iterative method to second order in a_0 and we may try to evaluate higher-order terms as a series expansion. We will not do this for an electron in a monochromatic wave since an *exact* solution exists (Sect. 2.1.3). Nevertheless it is worth to discuss for a moment other contributions to the second-order approximation. First, we have seen that when going beyond the linear approximation there is a displacement of the electron along the x axis. If we were to evaluate the fields at the actual particle position, we would find

$$\mathbf{E} = \mathbf{E}(\mathbf{r}(t), t) \simeq \mathbf{E}(x^{(1)} + x^{(2)}(t) + \cdots, t), \qquad (2.12)$$

so that keeping only the constant term $x^{(1)}$ in the argument, we have neglected a term $\sim x^{(2)}\partial_x \mathbf{E}(x^{(1)}) \sim x^{(2)} k \mathbf{E}(x^{(1)}) \sim a_0^3$, consistently with our approximation. Second, as a_0 approaches unity the motion becomes relativistic, thus we should use the relativistic equations for the momentum

$$\frac{d\mathbf{p}}{dt} = -e\left(\mathbf{E} + \frac{\mathbf{v}}{c} \times \mathbf{B}\right), \qquad \mathbf{p} = m_e \gamma \mathbf{v}, \qquad (2.13)$$

where $\gamma = (1 - v^2/c^2)^{-1/2} = 1 + v^2/2c^2 + \cdots$, so that $\mathbf{p} = m_e \mathbf{v} + \mathcal{O}(v^3/c^3)$, that is again consistent with keeping only the nonlinear $\mathbf{v} \times \mathbf{B}$ term for the first step in the approximation.

2.1.2 Relativistic Regime

The dimensionless quantity a_0 defined by (2.4) is the peak momentum of an electron oscillating in an electric field of frequency ω and amplitude E_0, in units of $m_e c$. As we infer from the preceding Sect. 2.1.1, where a_0 served as the expansion parameter for the approximate non-relativistic solution, a_0 is a convenient measure of the importance of relativistic effects: the "relativistic regime" of laser-plasma interaction may be defined by the condition $a_0 \gtrsim 1$. It is thus useful to relate a_0 to practical parameters such as the wavelength $\lambda = 2\pi c/\omega$ and the intensity I, defined as the cycle-averaged value of the modulus of the Poynting vector $\mathbf{S}$

$$I = \langle |\mathbf{S}| \rangle = \left\langle \frac{c}{4\pi} |\mathbf{E} \times \mathbf{B}| \right\rangle = \frac{c}{8\pi} |E_0|^2 = \frac{c}{8\pi} \left(\frac{m_e \omega c a_0}{e}\right)^2, \qquad (2.14)$$

so that we obtain in practical units[2]

---

[2] Notice that, in agreement with the definition (2.1), for a wave of given irradiance $I\lambda^2$ the *peak* dimensionless amplitude of the electric field is a_0 for linear polarization and $a_0/\sqrt{2}$ for circular polarization, with a_0 given by (2.15).

$$a_0 = 0.85 \left(\frac{I\lambda^2}{10^{18}\,\text{W cm}^{-2}} \right)^{1/2}. \tag{2.15}$$

Present-day laser systems allow to reach focused intensities above $10^{21}\,\text{W cm}^{-2}$ (Yanovsky et al. 2008) for $\lambda = 0.8\,\mu\text{m}$ (Ti:Sapphire lasers), so that strongly relativistic regimes with $a_0 \sim 10$ are at the forefront of current research. The relativistic regime is also accessible with CO_2 lasers ($\lambda \simeq 10\,\mu\text{m}$) currently reaching peak intensities $I \simeq 10^{16}\,\text{W cm}^{-2}$ (Haberberger et al. 2010).

2.1.3 Relativistic Motion in a Plane Wave

We now search for *exact* solutions of the relativistic motion of an electron in a plane wave EM field of arbitrary amplitude. The electron dynamics is thus described by Eq. (2.13) to which we add the equation for the electron energy,

$$\frac{d}{dt}(m_e \gamma c^2) = -e\mathbf{v} \cdot \mathbf{E}. \tag{2.16}$$

Our presentation closely follows that of Gibbon (2005) (see also Landau and Lifshitz 1975, Sects. 47, 48).

The EM plane wave, which is assumed to propagate long $\hat{\mathbf{x}}$, can be represented by the vector potential $\mathbf{A} = \mathbf{A}(x, t)$ with $\mathbf{A} \cdot \hat{\mathbf{x}} = 0$. The electric and magnetic fields are given by

$$\mathbf{E} = -\frac{1}{c}\partial_t \mathbf{A}, \qquad \mathbf{B} = \nabla \times \mathbf{A} = \hat{\mathbf{x}} \times \partial_x \mathbf{A}. \tag{2.17}$$

We let the "$\perp$" subscript refer to the vector components of fields and momentum in the transverse yz plane, e.g. $\mathbf{p}_\perp = (p_y, p_z)$. By noticing that $\mathbf{A} = \mathbf{A}_\perp$ and $(\mathbf{v} \times \mathbf{B})_\perp = -v_x \partial_x \mathbf{A}$, we obtain from (2.13)

$$\frac{d\mathbf{p}_\perp}{dt} = +\frac{e}{c}(\partial_t \mathbf{A} + v_x \partial_x \mathbf{A}) = +\frac{e}{c}\frac{d\mathbf{A}}{dt}, \tag{2.18}$$

which yields the following relation

$$\frac{d}{dt}\left(\mathbf{p}_\perp - \frac{e}{c}\mathbf{A}\right) = 0, \tag{2.19}$$

stating the conservation of canonical momentum due to translational invariance in the transverse plane.

By substituting $\mathbf{A}$ in the x-component of (2.13) and in (2.16) we obtain

$$\frac{dp_x}{dt} = -\frac{e}{c}(v_y \partial_x A_y + v_z \partial_x A_z), \qquad \frac{d}{dt}(m_e \gamma c) = \frac{e}{c^2}(v_y \partial_t A_y + v_z \partial_t A_z), \tag{2.20}$$

and by summing and subtracting these two equations we obtain

$$\frac{d}{dt}(p_x \mp m_e \gamma c) = -\frac{e}{c}\left[v_y \left(\partial_x \pm \frac{1}{c}\partial_t\right) A_y + v_z \left(\partial_x \pm \frac{1}{c}\partial_t\right) A_z\right]. \tag{2.21}$$

Let us now specify that $\mathbf{A}$ is a propagating plane wave, i.e. $\mathbf{A} = \mathbf{A}(x \mp ct)$ which are solutions of the uni-directional wave equation $(c\partial_x \pm \partial_t)\,\mathbf{A}(x \mp ct) = 0$. For definiteness we assume that the wave propagates in the positive direction of the x axis, i.e. $\mathbf{A} = \mathbf{A}(x - ct)$. Then we obtain

$$\frac{d}{dt}(p_x - m_e \gamma c) = 0 \tag{2.22}$$

which is a second conservation law. The trajectory of the electron is then determined implicitly by the values of the two constants of motion $\mathbf{p}_\perp - (e/c)\mathbf{A} \equiv \mathbf{p}_{\perp 0}$ and $m_e \gamma c - p_x \equiv -\alpha$.

To determine a possible choice for $\mathbf{p}_{\perp 0}$ and α, let us assume that the electron was initially at rest ($\mathbf{p} = 0$) and there was no field at that time ($\mathbf{A} = 0$). This corresponds to taking $\mathbf{p}_{\perp 0} = 0$ and $\alpha = -m_e c$. Physically, this choice corresponds to assuming that somewhere in the past the field must have been "turned on", which is indeed the case for a finite duration EM pulse whose front reaches at some time the electron initially at rest.

When $\mathbf{p}_{\perp 0} = 0$ and $\alpha = -m_e c$, we immediately obtain $\mathbf{p}_\perp = (e/c)\mathbf{A}$ and $p_x = mc(\gamma - 1)$. Using $\mathbf{p}^2 + m_e^2 c^2 = (m_e c \gamma)^2$ and $\mathbf{p}^2 = p_x^2 + \mathbf{p}_\perp^2$, with some simple algebra we obtain the relation

$$p_x = \frac{\mathbf{p}_\perp^2}{2m_e c} = \frac{1}{2m_e c}\left(\frac{e\mathbf{A}}{c}\right)^2. \tag{2.23}$$

The relation $p_x = mc(\gamma - 1)$ has an intuitive interpretation in terms of the energy and momentum delivered by the EM field. An EM plane wavepacket carrying the total energy U also contains the momentum $(U/c)\hat{\mathbf{x}}$ where $\hat{\mathbf{x}}$ is the propagation direction.[3] Thus, if the wave impinges on an electron that then acquires a kinetic energy $K = m_e c^2(\gamma - 1)$, a proportional amount of momentum $(K/c)\hat{\mathbf{x}}$ must have been also delivered, hence $p_x = K/c = m_e c(\gamma - 1)$.

Equations (2.19–2.23) relate directly the momentum components to the amplitude $\mathbf{A}$ of the EM wave. The actual trajectory is still implicit because $\mathbf{A}$ depends on the electron position $x(t)$. Nevertheless it is already evident that both p_x and $\mathbf{p}_\perp$ vanish when $\mathbf{A} = 0$, e.g. after the end of the EM pulse, thus the electron eventually absorbs no net energy from the wave. One may argue that after the pulse the fields $\mathbf{E} = \mathbf{B} = 0$

---

[3] Of course this property of the classical EM field corresponds to photons having energy $\hbar\omega$ and momentum $\hbar(\omega/c)\hat{\mathbf{x}}$ in the quantum theory of light. An electron may absorb an integer number of photons, hence the proportionality between kinetic energy and momentum is conserved.

necessarily but the "auxiliary" potential $\mathbf{A}$ is not necessarily zero. However, since $\mathbf{E} = -\partial_t \mathbf{A}/c$, the final value of $\mathbf{A}$ is related to the electric field by

$$\mathbf{A}(+\infty) = -c \int_{-\infty}^{+\infty} \mathbf{E}(\tau)d\tau, \tag{2.24}$$

where the integral gives the average value of $\mathbf{E}$, which is also the zero frequency component of the Fourier spectrum. Such component would not correspond to a propagating monochromatic wave. We may thus take $\mathbf{A}(+\infty) = \mathbf{A}(-\infty) = 0$ for a real, propagating plane wave EM pulse.

The conclusion that the electron gains no energy appears to be in agreement with the so-called Lawson-Woodward theorem, stating that a charge cannot be accelerated by a radiation field in vacuum extending over an infinite region (as it is the case for a plane wave). However, electron acceleration in vacuum is indeed possible by violating one or more hypotheses of the "no acceleration" theorem, as discussed by Gibbon (2005, Sect. 3.4). For instance, a very natural possibility is to use a focused laser pulse, and in such case under suitable assumptions the electron acceleration may be well described using the concept of the ponderomotive force, see Sect. 2.1.4. It may be of interest to notice that the problem of electron acceleration by EM pulses in vacuum is a long-standing and highly controversial issue; for example the reader may refer to several critical comments and replies published in the literature (McDonald 1998; Mora and Quesnel 1998; Kim et al. 2000; Troha and Hartemann 2002 and references therein).

We now focus on the case of a monochromatic wave of frequency ω. For the moment we still take for the constants of motion $\mathbf{p}_{\perp 0} = 0$ and $\alpha = -m_e c$, i.e. the choice corresponding to $\mathbf{A} = 0$ and $\mathbf{p} = 0$ in the remote past. This choice might sound contradictory because a perfectly monochromatic wave has an infinite duration and the field may never be assumed to vanish. In such a case, we assume the turn-on time to be arbitrarily long ("adiabatic rising"), so that the amplitude of the field can be eventually taken to be constant.

Since our equations now are fully nonlinear it is better to switch back to the real representation of the fields. We take for the vector potential

$$\mathbf{A} = A_0[\hat{\mathbf{y}}\delta \cos\phi + \hat{\mathbf{z}}(1 - \delta^2)^{1/2}\sin\phi], \qquad \phi = kx - \omega t, \tag{2.25}$$

where $\delta \leq 1$ is a parameter defining the polarization of the wave: for $\delta = 1$ or 0 the wave is linearly polarized along $\hat{\mathbf{y}}$ or $\hat{\mathbf{z}}$, respectively, while for $\delta = \pm 1/\sqrt{2}$ it is circularly polarized, and other values correspond to elliptical polarization.

When inserting (2.25) in the equations of motion, the phase ϕ must be considered a function of the particle position, i.e. $\phi = \phi[x(t), t]$. Since it will be convenient to use ϕ as the only variable we evaluate its derivative with respect to time as

$$\frac{d\phi}{dt} = \partial_t\phi + v_x\partial_x\phi = -\omega + \frac{p_x}{m_e\gamma}k = -\omega + \frac{m_e c(\gamma - 1)}{m_e\gamma}\frac{\omega}{c} = -\frac{\omega}{\gamma}. \tag{2.26}$$

Thus we also obtain

$$\mathbf{p} = m_e \gamma \frac{d\mathbf{r}}{dt} = m_e \gamma \frac{d\phi}{dt} \frac{d\mathbf{r}}{d\phi} = -m_e \omega \frac{d\mathbf{r}}{d\phi}, \qquad (2.27)$$

a very useful relation to obtain the particle trajectory. From (2.19) and (2.23) we immediately get

$$\mathbf{p}_\perp = (p_y, p_z) = \frac{eA_0}{c}(\delta \cos \phi, (1 - \delta^2)^{1/2} \sin \phi), \qquad (2.28)$$

$$p_x = \frac{1}{2m_e c} \left(\frac{eA_0}{c}\right)^2 [\delta^2 \cos^2 \phi + (1 - \delta^2) \sin^2 \phi]$$

$$= \frac{1}{4m_e c} \left(\frac{eA_0}{c}\right)^2 [1 + (2\delta^2 - 1) \cos 2\phi]. \qquad (2.29)$$

The term $\cos 2\phi$ is rapidly oscillating at the frequency 2ω (and higher frequencies, depending on $x(t)$), so that averaging over an oscillation cycle $\langle \cos 2\phi \rangle = 0$ and

$$\langle p_x \rangle = \frac{1}{4m_e c} \left(\frac{eA_0}{c}\right)^2 = m_e c \frac{a_0^2}{4} \equiv p_d. \qquad (2.30)$$

We thus have a constant drift of the electron. Furthermore we notice that for *circular* polarization, $2\delta^2 - 1 = 0$ and $p_x = \langle p_x \rangle$, i.e. there is no high frequency component in the longitudinal motion. Also notice that in this case $\mathbf{p}_\perp^2 = (eA_0/c)^2/2$ is a constant too, so that $\mathbf{p}^2 = \mathbf{p}_\perp^2 + p_x^2$ and the relativistic factor γ are constant as well.

We now use Eq. (2.27) to obtain the trajectory in implicit form by integrating (2.28, 2.29) with respect to ϕ:

$$x = x(\phi) = \frac{c}{\omega} \left(\frac{eA_0}{2m_e c^2}\right)^2 \left[-\phi - \left(\delta^2 - \frac{1}{2}\right) \sin 2\phi\right],$$

$$y = y(\phi) = -\frac{c}{\omega} \frac{eA_0}{m_e c^2} \delta \sin \phi, \quad z = z(\phi) = \frac{c}{\omega} \frac{eA_0}{m_e c^2} (1 - \delta^2)^{1/2} \cos \phi. \qquad (2.31)$$

By averaging over one cycle the first of these equations we obtain $k \langle x \rangle = -(a_0^2/4)k(\langle x \rangle - ct)$, being $eA_0/m_e c^2 = a_0$. We thus obtain the drift velocity v_d:

$$\langle x \rangle = \frac{a_0^2}{a_0^2 + 4} ct \equiv v_d t, \quad v_d = \frac{a_0^2}{a_0^2 + 4} c. \qquad (2.32)$$

To show the self-similar properties of the trajectory, we rewrite the above equations in dimensionless units, normalizing the coordinates to $1/k = c/\omega$. By defining $\hat{\mathbf{r}} = k\mathbf{r}$ (so that $\phi = \hat{x} - \hat{t}$ where $\hat{t} = \omega t$) we obtain

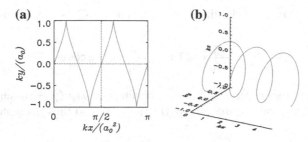

Fig. 2.2 a The "universal" trajectory of an electron in a monochromatic, linearly polarized wave plane wave of dimensionless amplitude a_0. **b** The trajectory in a circularly polarized plane wave for $a_0 = 2$

$$\frac{\hat{x}}{a_0^2} = \frac{1}{4}\left[-\phi - \left(\delta^2 - \frac{1}{2}\right)\sin 2\phi\right],$$

$$\frac{\hat{y}}{a_0} = -\delta\sin\phi, \qquad \frac{\hat{z}}{a_0} = (1-\delta^2)^{1/2}\cos\phi. \tag{2.33}$$

If the parameter a_0 changes, i.e. $a_0 \to \eta a_0$, the new orbit is obtained from the old one by stretching the coordinates according to $(\hat{x}, \hat{y}, \hat{z}) \to (\eta^2\hat{x}, \eta\hat{y}, \eta\hat{z})$. In other words, the trajectory can be represented as an *universal* curve is the reduced variables $\hat{x}/a_0^2$, $\hat{y}/a_0$ and $\hat{z}/a_0$ are used. To give an example, assuming linear polarization along $\hat{y}$ ($\delta = 1$), the universal curve is shown in Fig. 2.2.

The trajectory can be written out explicitly in a very simple way for circular polarization, since $\hat{x}/a_0^2 = -\phi/4 = -(\hat{x} - \hat{t})/4$. Thus, $\hat{x} = (v_d/c)\hat{t}$, while in the transverse plane $\hat{y} = (a_0/\sqrt{2})\sin\phi$, $\hat{z} = -(a_0/\sqrt{2})\cos\phi$, so that $\hat{y}^2 + \hat{z}^2 = a_0^2/2$. The trajectory is helicoidal (Fig. 2.2), with radius $R = a_0/\sqrt{2}k = a_0\lambda/(2\pi\sqrt{2})$ in the (y, z) plane.

Since the electron drifts at a constant velocity along x, transforming to a reference frame moving with a velocity $V_x = v_d$ we obtain in such frame that for circular polarization the orbit is a closed circle. Also for linear polarization the electron orbit is closed in the frame where the average velocity vanishes. Such orbit is the already seen "figure of eight" as we now show.

We look for the frame where $\langle p_x \rangle = 0$. This is equivalent to a different choice of the constant of motion $\alpha = p_x - m_e\gamma c$. In fact, using $\gamma = \sqrt{\mathbf{p}^2/m_e^2c^2 + 1}$, eliminating γ we obtain with some simple algebra

$$p_x = \frac{\mathbf{p}_\perp^2 + m_e^2c^2 - \alpha^2}{2\alpha} = \frac{e^2\mathbf{A}^2/c^2 + m_e^2c^2 - \alpha^2}{2\alpha}. \tag{2.34}$$

Thus, posing $\langle p_x \rangle = 0$ we obtain

$$\alpha^2 = m_e^2c^2 + \left(\frac{e}{c}\right)^2\langle\mathbf{A}^2\rangle = m_e^2c^2 + \left(\frac{e}{c}\right)^2\frac{A_0^2}{2} \equiv m_e^2c^2\gamma_0^2, \tag{2.35}$$

where $\gamma_0^2 \equiv (1 + a_0^2/2)^{1/2}$. Putting $\alpha = m_e c \gamma_0$ back in (2.34), we obtain for $\delta = 1$

$$\frac{p_x}{m_e c} = \frac{a_0^2}{4\gamma_0} \cos 2\phi, \qquad \frac{p_y}{m_e c} = a_0 \cos \phi. \qquad (2.36)$$

Now notice that, from $m_e \gamma c = p_x + \alpha$ we obtain $p_x = m_e c (\gamma - \gamma_0)$ and $d\phi/dt = -(\gamma_0/\gamma)\omega$, so that now $d\mathbf{r}/d\phi = -\mathbf{p}/(m_e \gamma_0 \omega)$. We can thus integrate over ϕ to obtain

$$\hat{x} = \frac{a_0^2}{8\gamma_0} \sin 2\phi, \qquad \hat{y} = -\frac{a_0}{\gamma_0} \sin \phi, \qquad (2.37)$$

which corresponds again to a trajectory of the form (2.10) as shown in Fig. 2.1 where now $X = (\omega x/c)\gamma_0/a_0^2$ and $Y = (\omega y/c)\gamma_0/a_0$. The non-relativistic case is recovered by posing $\gamma_0 = 1$.

2.1.4 Ponderomotive Force

The study of the motion in a monochromatic plane wave is an instructive example of relativistic electron dynamics. The availability of an exact solution to the problem can be useful for numerical benchmarks.

However, "realistic" electromagnetic fields or, to be more specific, laser pulses are not perfectly monochromatic plane waves, but have finite width and duration. In general, a laser pulse may be described by an envelope function, describing its transverse and longitudinal profiles, multiplied by an oscillating function:

$$\mathbf{E}(\mathbf{r}, t) = \text{Re}\left(\tilde{\mathbf{E}}(\mathbf{r}, t) e^{-i\omega t} \right) = \frac{1}{2} \tilde{\mathbf{E}}(\mathbf{r}, t) e^{-i\omega t} + \text{c.c.},$$

$$\mathbf{B}(\mathbf{r}, t) = \text{Re}\left(\tilde{\mathbf{B}}(\mathbf{r}, t) e^{-i\omega t} \right) = \frac{1}{2} \tilde{\mathbf{B}}(\mathbf{r}, t) e^{-i\omega t} + \text{c.c.}. \qquad (2.38)$$

The envelope functions $\tilde{\mathbf{E}}$ and $\tilde{\mathbf{B}}$ are supposed to vary with time on a scale slower than the oscillation period $T = 2\pi/\omega$. We assume that the field (almost) averages to zero over a period, i.e. $\langle \mathbf{E}(\mathbf{r}, t) \rangle \simeq 0$, while for the envelope function $\left\langle \tilde{\mathbf{E}}(\mathbf{r}, t) \right\rangle \neq 0$. The assumption of two separate time scales, a slow and a fast one, suggests us to describe the electron motion as the superposition of a slow term and a fast ("oscillating") term:

$$\mathbf{r}(t) = \mathbf{r}_s(t) + \mathbf{r}_o(t), \qquad \langle \mathbf{r}_o(t) \rangle = 0, \qquad \langle \mathbf{r}(t) \rangle = \langle \mathbf{r}_s(t) \rangle = \mathbf{r}_s(t). \qquad (2.39)$$

It is possible, under suitable conditions, to describe the "slow motion" by a dynamic equation with a slowly-varying force which is named the ponderomotive force (PF).

In other words, we study the motion of an "oscillation center" over which a fast oscillation is superposed.[4]

We derive the PF in the non-relativistic regime, keeping terms up to order $\sim v/c \ll 1$, and discuss the relativistic regime later. A further and crucial assumption is that across an oscillation the spatial variation of the field envelope is small. Since the oscillation amplitude is less than $cT = \lambda$, it is sufficient that the scale of spatial variation of $\tilde{\mathbf{E}}$ is quite larger than λ, which is appropriate in many cases. This assumption allows to expand the field as follows

$$\mathbf{E}(\mathbf{r}(t), t) = \mathbf{E}(\mathbf{r}_s(t) + \mathbf{r}_o(t), t) \simeq \mathbf{E}(\mathbf{r}_s(t), t) + (\mathbf{r}_o(t) \cdot \nabla)\mathbf{E}(\mathbf{r}_s(t), t). \quad (2.40)$$

To lowest order, the equations for the oscillating components are thus

$$\frac{d^2\mathbf{r}_o}{dt^2} = \frac{d\mathbf{v}_o}{dt} \simeq -\frac{e}{m_e}\mathbf{E}(\mathbf{r}_s(t), t) \simeq \frac{-e}{2m_e}\tilde{\mathbf{E}}(\mathbf{r}_s(t))e^{-i\omega t} + \text{c.c.}, \quad (2.41)$$

so that $\mathbf{r}_o = \text{Re}\left(\tilde{\mathbf{r}}_o e^{-i\omega t}\right)$ and $\mathbf{v}_o = \text{Re}\left(\tilde{\mathbf{v}}_o e^{-i\omega t}\right)$ with

$$\tilde{\mathbf{r}}_o = \frac{e}{m_e\omega^2}\tilde{\mathbf{E}}(\mathbf{r}_s(t)), \qquad \tilde{\mathbf{v}}_o = -\frac{ie}{m_e\omega}\tilde{\mathbf{E}}(\mathbf{r}_s(t)). \quad (2.42)$$

Now, by averaging Newton's equation we have

$$m_e\frac{d\mathbf{v}_s}{dt} = -e\langle\mathbf{E}(\mathbf{r}(t), t)\rangle - \frac{e}{c}\langle\mathbf{v} \times \mathbf{B}(\mathbf{r}(t), t)\rangle. \quad (2.43)$$

For the electric field average we have[5]

$$\langle\mathbf{E}(\mathbf{r}(t), t)\rangle \simeq \langle\mathbf{E}(\mathbf{r}_s(t), t) + (\mathbf{r}_o(t) \cdot \nabla)\mathbf{E}(\mathbf{r}_s(t), t)\rangle$$

$$= \frac{1}{4}(\tilde{\mathbf{r}}_o^*(t) \cdot \nabla)\tilde{\mathbf{E}}(\mathbf{r}_s(t), t) + \text{c.c.}$$

$$= \frac{e}{4m_e\omega^2}(\tilde{\mathbf{E}}^*(\mathbf{r}_s(t), t) \cdot \nabla)\tilde{\mathbf{E}}(\mathbf{r}_s(t), t) + \text{c.c.}. \quad (2.44)$$

For the magnetic force term we have

---

[4] This is analogous to the "guiding center" approximation to study the motion in an inhomogeneous magnetic field, in which a charged particle drifts slowly along the field lines while rotating around them with the local cyclotron frequency.

[5] Notice that if $A(t) = \text{Re}(\tilde{A}e^{-i\omega t})$ and $B(t) = \text{Re}(\tilde{B}e^{-i\omega t})$, where $\tilde{A}$ and $\tilde{B}$ are slowly-varying functions of time, then $\langle A(t)B(t)\rangle = \text{Re}(\tilde{A}\tilde{B}^*)/2$. In particular, $\langle A^2(t)\rangle = |\tilde{A}^2|/2$.

$$\langle \mathbf{v} \times \mathbf{B}(\mathbf{r}(t), t) \rangle \simeq \frac{1}{4} \tilde{\mathbf{v}}_0^* \times \tilde{\mathbf{B}}_o(\mathbf{r}_s(t), t) + \text{c.c.}$$

$$= -\frac{ec}{4m_e \omega^2} \tilde{\mathbf{E}}^*(\mathbf{r}_s(t), t) \times \left(\nabla \times \tilde{\mathbf{E}}(\mathbf{r}_s(t), t) \right) + \text{c.c.}, \qquad (2.45)$$

where we used $c\nabla \times \mathbf{E} = -\partial_t \mathbf{B}$ i.e. $c\nabla \times \tilde{\mathbf{E}} = (\mathrm{i}\omega)\tilde{\mathbf{B}}$. We thus obtain

$$m_e \frac{d\mathbf{v}_s}{dt} \simeq -\frac{e^2}{4m_e \omega^2} \left(\tilde{\mathbf{E}}^*(\mathbf{r}_s(t), t) \cdot \nabla \right) \tilde{\mathbf{E}}(\mathbf{r}_s(t), t)$$

$$-\tilde{\mathbf{E}}^*(\mathbf{r}_s(t), t) \times \nabla \times \tilde{\mathbf{E}}(\mathbf{r}_s(t), t) \Big) + \text{c.c.}$$

$$= -\frac{e^2}{4m_e \omega^2} \nabla |\mathbf{E}^*(\mathbf{r}_s(t), t)|^2 = -\frac{e^2}{2m_e \omega^2} \nabla \left\langle \mathbf{E}^2(\mathbf{r}_s(t), t) \right\rangle \equiv \mathbf{f}_p. \quad (2.46)$$

The latter equality defines the PF $\mathbf{f}_p$, which describes the dynamics of the oscillation center, i.e. of the cycle-averaged position and velocity according to

$$m_e \frac{d^2 \langle \mathbf{r} \rangle}{dt^2} = m_e \frac{d \langle \mathbf{v} \rangle}{dt} = \mathbf{f}_p = -\nabla \Phi_p, \qquad (2.47)$$

where the so-called "ponderomotive potential" Φ_p is actually the cycle-averaged oscillation energy, which is assumed to be a function of the oscillation center position

$$\Phi_p = \Phi_p(\langle \mathbf{r} \rangle) = \frac{m_e}{2} \left\langle \mathbf{v}_o^2 \right\rangle = \frac{e^2}{2m_e \omega^2} \left\langle \mathbf{E}^2 \right\rangle. \qquad (2.48)$$

The most important consequence of (2.47, 2.48) is that electrons will be scattered off, i.e. expelled from regions where the electric field is higher. This is actually true also for positively charged particles, since $\mathbf{f}_p$ is proportional to the charge squared. However, $\mathbf{f}_p$ also scales with the inverse of the particle mass, thus in general the ponderomotive effect on ions is negligible with respect to that on free electrons.

While the PF description is useful and appropriate in many situations, one should keep in mind that the approximations underlying its derivation break down in certain regimes. One prominent example is that of a laser pulse which is not much longer or wider than λ, i.e. a few-cycle or a tightly focused pulse; both conditions are experimentally feasible nowadays.

To provide an example of application of (2.46), let us consider an electron which is initially at rest in the field of an EM plane wave whose intensity rises up over a long time, and then remains constant. As discussed in Sect. 2.1.3 a constant drift velocity should then appear. If we use the PF to describe the longitudinal motion, we have the equation

$$m_e \frac{dv_s}{dt} = -\frac{m_e c^2}{2} \partial_x \left\langle a^2 \right\rangle, \qquad (2.49)$$

where $a = a(x - ct)$ represents the traveling wave. Now, using the variable $\zeta = x_s(t) - ct$, we have $d_t \zeta = v_s - c$, $\partial_x = \partial_\zeta$, and $d_t = (d_t \zeta) d_\zeta$, thus we rewrite (2.49) as

$$m_e(v_s - c)\frac{dv_s}{d\zeta} = -\frac{m_e c^2}{2}\frac{d \langle a^2(\zeta) \rangle}{d\zeta}, \qquad (2.50)$$

that can be integrated from the initial time, when $v_s = 0$ and $a = 0$, to the late time when $a = a_0$ and $\langle a \rangle^2 = a_0^2/2$, obtaining $v_s^2/2 - v_s c = -a_0^2 c^2/4$ which for $a_0 \ll 1$ gives $v_s \simeq a_0^2 c/4$ in agreement with (2.30).

A theory of the PF in the relativistic regime is not trivial and has been a subject of considerable controversy. The reader may look at Mulser and Bauer (2010), Chap. 5, for a more extended discussion and references. Here we only mention that, in the case of electromagnetic waves, it turns out that $\mathbf{f}_p$ is again minus the gradient of the cycle-averaged oscillation energy. For the motion in an EM in a laser pulse of finite extension (both in length and width) described by the vector potential $\mathbf{A}(\mathbf{r}, t)$, it can be shown that $\mathbf{p}_\perp \simeq (e/c)\mathbf{A}$ approximately holds for a non-plane wave if the intensity does not vary too much over a distance $\simeq \lambda$ (see also Sect. 3.3.2). Supposing that $|\mathbf{p}| \simeq |\mathbf{p}_\perp|$, then the kinetic energy $m_e(\gamma - 1)c^2 \simeq m_e c^2((1 + \mathbf{a}^2)^{1/2} - 1)$ with $\mathbf{a} = e\mathbf{A}/m_e c^2$. In these conditions

$$\mathbf{f}_p = -m_e c^2 \nabla \left(1 + \langle \mathbf{a}^2 \rangle\right)^{1/2} = -\nabla m_{\text{eff}} c^2, \qquad m_{\text{eff}} \equiv m_e \left(1 + \langle \mathbf{a}^2 \rangle\right)^{1/2}. \quad (2.51)$$

The "effective mass" m_{eff} has been introduced to point out that the oscillating momentum leads to an increase of the effective inertia, which has to be taken into account: thus, the equation for the "slow" motion of the oscillating center, in the case in which the cycle-averaged velocity is small with respect to c, can be written as

$$\frac{d}{dt}(m_{\text{eff}}\mathbf{v}_s) = -\nabla m_{\text{eff}} c^2. \qquad (2.52)$$

2.1.5 Radiation Friction

In the preceding sections, we studied the dynamics of an electron in an external EM field considered as given and independent of the electron motion. From the knowledge of the latter one may calculate in turn the EM radiation emitted by the accelerated electron. This approach, however, is not consistent because in principle both the external and the radiation field act on the electron. This issue brings us to the complicated problem of the back-action or *reaction* on the electron by the EM field radiated by the electron itself. As a consequence of such reaction, the electron dynamics cannot be described by the Lorentz force alone anymore, and an additional force term $\mathbf{f}_{\text{rad}}$ appears:

$$\frac{d\mathbf{p}}{dt} = -e\left(\mathbf{E} + \frac{\mathbf{v}}{c} \times \mathbf{B}\right) + \mathbf{f}_{\text{rad}}. \tag{2.53}$$

To justify the need to include $\mathbf{f}_{\text{rad}}$ in the electron dynamics and its name as the *radiation friction* (RF) force[6] the following example may be useful. Consider and electron in an external field, such that electron is accelerated but its total mechanical energy is a constant of motion (when radiation is neglected). The simplest case is that of an electron (non-relativistic for simplicity) having velocity $\mathbf{v}$ in an external magnetic field $\mathbf{B}_0$, although we could also consider an electron bound by an harmonic or Keplerian potential. Taking $\mathbf{v}$ as perpendicular to $\mathbf{B}_0$, from the solution of the equation of motion $m_e d\mathbf{v}/dt = -e\mathbf{v} \times \mathbf{B}_0/c$ it is readily found that the electron orbits along a circle of radius $R_L = v/\omega_c$ where $\omega_c = eB_0/m_e c$. However, when we turn the radiation on, the electron will radiate a total power given by Larmor's formula as

$$P_{\text{rad}} = \frac{2e^2}{3c^3}|\dot{\mathbf{v}}|^2 = \frac{2e^4 B_0^2}{3m_e^2 c^5}v^2 = \frac{2e^2 \omega_c^2}{3c^3}v^2, \tag{2.54}$$

where, following the usual notation, the upper dot denotes a derivative with respect to time. The power loss must be at the expense of the electron's energy $K = m_e v^2/2$, which will then decrease according to $dK/dt = -P_{\text{rad}}$, i.e.

$$m_e v \frac{dv}{dt} = -\frac{2e^2 \omega_c^2}{3c^3}v^2, \qquad v(t) = v(0)\exp(-t/\tau), \tag{2.55}$$

where $\tau = 3m_e c^3/(2e^2 \omega_c^2) = 3c/(2r_c \omega_c^2)$. Since ω_c is constant, also the orbit radius $R_L = v/\omega_c$ will decrease and the electron will move along a spiral rather than remaining on circle, as if there was a friction force acting on the electron. Although this motion is physically sound, and a very reasonable approximation if the energy damping time is much larger than one period ($\tau \gg 2\pi/\omega_c$), it is clear that for a correct description of the dynamics a suitable expression for $\mathbf{f}_{\text{rad}}$ needs to be included in Eq. (2.53), since neither the trajectory nor the radiated power are calculated consistently when only the Lorentz force is kept. The idea is thus to determine $\mathbf{f}_{\text{rad}}$ in order that the work it does over some interval of time equals the radiated power:

$$\int_0^t \mathbf{f}_{\text{rad}} \cdot \mathbf{v}dt := -\int_0^t \frac{2e^2}{3c^3}|\dot{\mathbf{v}}|^2 \, dt = \frac{2e^2}{3c^3}\left(\int_0^t \ddot{\mathbf{v}} \cdot \mathbf{v}dt - \dot{\mathbf{v}} \cdot \mathbf{v}|_0^t\right). \tag{2.56}$$

If the last term vanishes, as for a periodic motion, we obtain

$$\mathbf{f}_{\text{rad}} = \frac{2e^2}{3c^3}\ddot{\mathbf{v}}. \tag{2.57}$$

---

[6] Often also the name *radiation reaction* force is used.

This expression is however unsatisfactory because of both the need of an additional initial condition (i.e. the initial value of the acceleration $\dot{\mathbf{v}}(0)$) to determine the solution of the equation of motion and the possibility of unphysical *runaway* solutions in the absence of an external force:

$$\mathbf{a}(t) = \dot{\mathbf{v}}(t) = \mathbf{a}(0)e^{t/\tau}, \qquad \tau \equiv 2e^2/(3m_e c^3). \tag{2.58}$$

This is just the starting point of a long standing, non trivial, and still not completely solved problem that ultimately deals with the fundamental nature of the electron (Jackson 1998, Sect. 16.1). The possibly good news are that RF effects are important only for extreme accelerations, hence for very intense fields, and ultrarelativistic electrons, thus in most situations $\mathbf{f}_{rad}$ is negligible in the electron dynamics and the not self-consistent calculation of the emitted radiation is accurate enough.

The situation becomes different with superintense lasers which may create conditions in which RF effects are non negligible. This perspective has greatly revitalized the interest in RF, because of the possibility to test directly the theory in laboratory experiments. Moreover, on the way towards the study of QED effects or other High Field Science issues with lasers, the higher the intensity the more important RF effects become, and thus they need to be included in theoretical models.

Much recent theoretical work has indeed focused on derivations of $\mathbf{f}_{rad}$ possibly more accurate or rigorous than previous calculations from the era preceding the advent of superintense lasers. As our aim is to remain at a textbook level, here we do not discuss most recent suggestions but we focus on the approach described by Landau and Lifshitz in their classic book (Landau and Lifshitz 1975, Sect. 76). The idea is to adopt an iterative approach starting from Eq. (2.57): in computing the first derivative $\dot{\mathbf{v}}$, only the Lorentz force is considered, so that $\mathbf{f}_{rad}$ becomes proportional to a total time derivative of the fields. Thus, from Eq. (2.57) we obtain

$$\mathbf{f}_{rad} \simeq -\frac{2e^3}{3m_e c^3}\frac{d}{dt}\left(\mathbf{E} + \frac{\mathbf{v}}{c}\times\mathbf{B}\right) \simeq -\frac{2e^3}{3m_e c^3}\left(\dot{\mathbf{E}} - \frac{e}{m_e c}\mathbf{E}\times\mathbf{B}\right), \tag{2.59}$$

where for simplicity we assumed $\mathbf{v} = 0$ (but $\dot{\mathbf{v}} \neq 0$), i.e. the expression is valid in the system where the electron is instantaneously at rest. Evidently, this expression is free from the problems of additional initial conditions and runaway solutions, since it does not contain time derivatives of $\dot{v}$ and it vanishes for zero fields. The approximation is justified if $|\mathbf{f}_{rad}| \ll |e\mathbf{E}|$. By considering a field with a typical frequency ω, such that $|\dot{\mathbf{E}}| \sim \omega E$, the conditions that both two terms in Eq. (2.59) are small with respect to eE are equivalent to

$$r_c = \frac{e^2}{m_e c^2} \ll \frac{c}{\omega}, \qquad eB \ll \frac{m_e^2 c^4}{e^2} = \frac{m_e c^2}{r_c}. \tag{2.60}$$

These conditions are always fulfilled within the limits of validity of classical electrodynamics. Notice, e.g., that $m_e c^2/(er_c) \simeq 6 \times 10^{15}$ G.

The above example, based on the non-relativistic expression, is pedagogically simple but of limited practical interest since the RF force is relevant only for strongly relativistic electrons. An analogous procedure using the relativistic expressions for the equation of motion and for the radiated power yields (Landau and Lifshitz 1975, Sect. 76)

$$
\mathbf{f}_{rad} = \frac{2r_c^2}{3} \left\{ -\gamma^2 \left[\left(\mathbf{E} + \frac{\mathbf{v}}{c} \times \mathbf{B} \right)^2 - \left(\frac{\mathbf{v}}{c} \cdot \mathbf{E} \right)^2 \right] \frac{\mathbf{v}}{c} \right.
$$
$$
\left. + \left[\left(\mathbf{E} + \frac{\mathbf{v}}{c} \times \mathbf{B} \right) \times \mathbf{B} + \left(\frac{\mathbf{v}}{c} \cdot \mathbf{E} \right) \mathbf{E} \right] - \gamma \frac{m_e c}{e} \left(\dot{\mathbf{E}} + \frac{\mathbf{v}}{c} \times \dot{\mathbf{B}} \right) \right\}, \quad (2.61)
$$

where $(\dot{\mathbf{E}}, \dot{\mathbf{B}}) = (\partial_t + \mathbf{v} \cdot \mathbf{\nabla})(\mathbf{E}, \mathbf{B})$. The first term in Eq. (2.61) is common to other proposed expressions for $\mathbf{f}_{rad}$ and is the dominant one (because of the $\sim \gamma^2$ scaling) in many situations of practical interest.

When the Landau–Lifshitz force (2.61) is added to the Lorentz force, an *exact* solution for the motion of an electron in a plane wave still can be found (Di Piazza 2008). Figure 2.3 shows the corresponding trajectory in the average rest frame of Fig. 2.1, showing that the figure-of-eight opens up and the electron is accelerated in this frame. Notice that acceleration is *not* "directly" due to $\mathbf{f}_{rad}$; the electron can now gain a net energy and momentum because the insertion of a friction force breaks the conservation of the invariants discussed in Sect. 2.1.3.

The exact solution in a plane wave is a very useful reference case, especially for testing the implementation of the RF force in laser-plasma simulation codes. At very high intensities, such inclusion is necessary because otherwise the energy and momentum loss due to high-frequency radiation would be neglected, possibly leading to a non-consistent dynamics. Examples of simulations of laser-plasma interactions

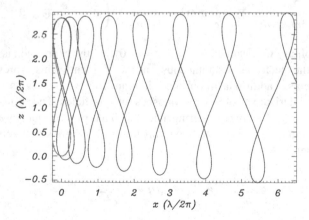

Fig. 2.3 The drifting "figure of eight" trajectory (*black line*) for an electron in a plane wave when the radiation friction force $\mathbf{f}_{rad}$ is included. The *red line* gives the usual closed trajectory obtained when $\mathbf{f}_{rad}$ is neglected. In this example, the wave amplitude $a_0 = 100$. Courtesy of M. Tamburini

at ultra-high intensity including RF effects may be found in the literature (Naumova, et al. 2009; Tamburini et al. 2010; Chen et al. 2011; Capdessus et al. 2012).

From (2.61) it also follows that if the electron is strongly relativistic with $v_x \simeq c$ and moves parallel to the wavevector $\mathbf{k} = k\hat{\mathbf{x}}$ of the plane wave, then $\mathbf{f}_{rad}$ vanishes for parallel propagation (since both $\mathbf{E} \cdot \mathbf{v} = 0$ and $\mathbf{E} + \mathbf{v} \times \mathbf{B}/c \rightarrow 0$) while it is maximized for anti-parallel, "colliding" propagation.[7] One may thus expect to observe strong signatures of RF in the scattering of an EM wave by a counterpropagating, highly relativistic electron beam (Di Piazza et al. 2009). For such experimental proposal laser-plasma electron accelerators (Sect. 4.1) may be very suitable for an "all-optical" experiment thanks to easy synchronization and the short duration and high density of the accelerated electron bunch. Notice that at present this experiment would bring a first direct test of RF.

A classical description of RF (as well as of classical electrodynamics in general) is not valid anymore when the electric field on an electron in its instantaneous rest frame approaches the *Schwinger field*

$$E_s = \frac{m_e c^2}{\lambda_c} = \frac{m_e^2 c^3}{e\hbar} = 1.3 \times 10^{16} \, \text{V cm}^{-1}, \tag{2.62}$$

where $\lambda_c = \hbar/m_e c$ is the Compton length. The value of E_s is a reference for nonlinear quantum electrodynamics (QED) effects such as pair production in vacuum, since for instance the work done by eE_s over a Compton length equals the rest energy of an electron. Since $E_s < m_e c^2/(e r_c) \simeq 2 \times 10^{18} \, \text{V cm}^{-1}$, the limitation on the validity of the classical RF force (2.61) comes from the onset of QED effects rather than from the breakdown of the ordering assumptions on which the iterative approach is based. The Schwinger field corresponds to an intensity of $2.3 \times 10^{29} \, \text{W cm}^{-2}$, still some eight orders of magnitude away from the values presently obtained in laboratory. However, in the rest frame of a relativistic electron the field may be much higher than in the laboratory. For instance, if the electron velocity $\mathbf{V} = \boldsymbol{\beta} c$ is anti-parallel to the wave vector of an EM wave of amplitude E_0, due to Lorentz transformations the electric field in the electron frame $E_0' = \gamma(1 + \beta) \simeq 2\gamma E_0 \gg E_0$ when $\beta \rightarrow 1$. The wave frequency is also upshifted by the same factor, i.e. $\omega' = \gamma(1 + \beta)\omega$, and quantum effects would come into play when the energy of a photon becomes of the order of the electron rest energy, i.e. $\hbar\omega' = m_e c^2$ or $\omega' = c/\lambda_c$. A QED description of RF might be necessary in particular conditions to be met in future experiments at extreme intensities (see e.g. Di Piazza et al. 2010).

---

[7] It may be also noticed that $\mathbf{f}_{rad}$ vanishes in a constant and uniform electric field, although the electron actually emits radiation. A discussion of this subtle issue is given by Peierls (1979) Sect. 8.1.

2.2 Macroscopic Description

2.2.1 Kinetic Equations

The most complete description of a classical, collisionless plasma is based on the knowledge, for each particle species a, of the distribution function $f_a = f_a(\mathbf{r}, \mathbf{p}, t)$ which gives the density of particles at the point $(\mathbf{r}, \mathbf{p})$ in the six-dimensional phase space at the time t. For a collisionless system in which the number of particles is conserved for each species, f_a obeys the kinetic equation

$$\partial_t f_a + \nabla_{\mathbf{r}} \cdot (\dot{\mathbf{r}}_a f_a) + \nabla_{\mathbf{p}} \cdot (\dot{\mathbf{p}}_a f_a) = 0, \qquad (2.63)$$

where $\dot{\mathbf{r}}_a = \mathbf{v} = \mathbf{p}/(m_a \gamma_a) = \mathbf{p}c/(m_a^2 c^2 + \mathbf{p}^2)^{1/2}$, $\dot{\mathbf{p}}_a = \mathbf{F}_a$ with $\mathbf{F}_a = \mathbf{F}_a(\mathbf{r}, \mathbf{p}, t)$ the force on the particle, and the gradient operators $\nabla_{\mathbf{r}}$ and $\nabla_{\mathbf{p}}$ act on the space $(\mathbf{r})$ and momentum $(\mathbf{p})$ variables, respectively. Equation (2.63) can be understood as a continuity equation in the phase space.

The total density $n_a = n_a(\mathbf{r}, t)$ in coordinate space and mean velocity $\mathbf{u}_a = \mathbf{u}_a(\mathbf{r}, t)$ of the species a are obtained by integrating over the momentum space:

$$n_a = \int f_a d^3 p, \qquad \mathbf{u}_a = n_a^{-1} \int \mathbf{v} f_a d^3 p. \qquad (2.64)$$

The Vlasov theory of a plasma assumes that $\mathbf{F}_a$ is the Lorentz force $\mathbf{F}_L = q_a(\mathbf{E} + \mathbf{v} \times \mathbf{B}/c)$ and that the EM fields are obtained self-consistently via Maxwell's equations where the source terms, i.e. the charge density $\varrho = \varrho(\mathbf{r}, t)$ and the current density $\mathbf{J} = \mathbf{J}(\mathbf{r}, t)$, are obtained as follows using (2.64):

$$\varrho = \sum_a q_a n_a, \qquad \mathbf{J} = \sum_a q_a n_a \mathbf{u}_a, \qquad (2.65)$$

where the sum runs over all the particles species, i.e. electrons and ions. The resulting coupled, nonlinear equations constitute the so-called Vlasov–Maxwell system. Notice that $\nabla_{\mathbf{r}} \cdot (\mathbf{v} f_a) = \mathbf{v} \cdot \nabla_{\mathbf{r}} f_a$ and $\nabla_{\mathbf{p}} \cdot (\mathbf{F}_L f_a) = \mathbf{F}_L \cdot \nabla_{\mathbf{p}} f_a$, so that the standard Vlasov equation may be written as

$$\partial_t f_a + \mathbf{v} \cdot \nabla_{\mathbf{r}} f_a + q_a (\mathbf{E} + \mathbf{v} \times \mathbf{B}/c) \cdot \nabla_{\mathbf{p}} f_a = 0. \qquad (2.66)$$

However, there are extensions of the Vlasov–Maxwell description including additional forces in $\mathbf{F}_a$ for which $\nabla_{\mathbf{p}} \cdot (\mathbf{F}_a f_a) \neq \mathbf{F}_a \cdot \nabla_{\mathbf{p}} f_a$ may occur, an important example being the Landau–Lifshitz formula (2.61) for the radiation friction force. In such cases, the more general Eq. (2.63) must be used.

Analytical solutions to the Vlasov–Maxwell system are very hard to find, and describe a limited set of physical phenomena. Indeed, there are efficient numerical methods for the solution of the Vlasov–Maxwell system, the most widely used being the Particle-In-Cell (PIC) approach (Dawson 1983; Birdsall and Langdon 1991) that is briefly discussed in Sect. 2.3.2. Still, solving Eq. (2.63) numerically can be an exceptionally demanding computational task, mostly because of the six-dimensional nature of the phase space. Although the enormous development of supercomputers presently allows to perform PIC simulation in a "real" 3D coordinate space and to approach quite "realistic" spatial and temporal scales of experiments, there still is a great need of models providing suitable simplifications which can make a specific problem more tractable. Often, models are obtained exploiting the large difference in the typical temporal scale of electron versus ion dynamics, due to the very different mass. It is then sometimes possible, e.g., to relax the computational effort by assuming an electrostatic approximation for the ion dynamics and/or a ponderomotive approximation for the electron dynamics. It is also possible to keep a kinetic description for one species only and to assume for the other species a simpler description which might be based, e.g., on the fluid equations which are discussed in the next section. However, when details of the particle distribution are needed, as for example when simulating an experiment of laser-plasma acceleration with the aim to obtain detailed spectra of either electrons or ions, the effort to solve the full kinetic equations must be pursued.

2.2.2 Fluid Equations

Roughly speaking, the idea beneath the so-called fluid description of a plasma is to obtain from the Vlasov equation (2.66) a set of equations for the averaged quantities n_a and $\mathbf{u}_a$ defined by (2.64), within the assumption that general properties of the distribution function are known. Let us sketch the procedure starting from the non-relativistic case. Integrating (2.66) over momentum leads to the continuity equation

$$\partial_t n_a + \nabla \cdot (n_a \mathbf{u}_a) = 0. \tag{2.67}$$

Now we multiply (2.66) by $\mathbf{v}$ and again integrate over momentum. The first and the third term give

$$\int \mathbf{v} \partial_t f_a d^3 p = \partial_t (n_a \mathbf{u}_a), \tag{2.68}$$

$$\int \mathbf{v} \left(\mathbf{E} + \frac{\mathbf{v}}{c} \times \mathbf{B} \right) \cdot \nabla_{\mathbf{p}} f_a d^3 p = n_a \left(\mathbf{E} + \frac{\mathbf{u}_a}{c} \times \mathbf{B} \right), \tag{2.69}$$

having assumed that f_a vanishes rapidly as $p \rightarrow \infty$. To tackle the integration of the second term of (2.66), we write $\mathbf{v} = \mathbf{u}_a + \mathbf{w}$ where $\mathbf{w}$ can be considered as the thermal or "random" velocity component. Since the average of $\mathbf{w}$ over f_a vanishes, we obtain[8]

---

[8] Here repeated indices imply a summation, i.e. $a_i b_i$ stands for $\sum_i a_i b_i$.

$$\int \mathbf{v}\,(v_i \cdot \nabla_{\mathbf{r}} f_a)\,d^3 p = \frac{\partial}{\partial r_j} \int v_i v_j f_a d^3 p$$

$$= m_a \frac{\partial}{\partial r_j} \int (u_{a,i} + w_i)(u_{a,j} + w_j) f_a d^3 w$$

$$= m_a \frac{\partial}{\partial r_j} (n_a u_{a,i} u_{a,j}) + m_a \int w_i w_j f_a d^3 w. \qquad (2.70)$$

The last term is the pressure tensor $P_{a,ij}$. Assuming for simplicity an isotropic distribution of the thermal velocities, $P_{a,ij} = P_a \delta_{ij}$. After some algebraic manipulations and also using (2.67) we eventually obtain the non-relativistic fluid equation for $\mathbf{u}_a$ with a scalar pressure term

$$m_a n_a (\partial_t \mathbf{u}_a + \mathbf{u}_a \cdot \nabla \mathbf{u}_a) = q_a n_a \,(\mathbf{E} + \mathbf{u}_a \times \mathbf{B}/c) - \nabla P_a. \qquad (2.71)$$

In order to make (2.67) and (2.71) a closed system of equations (together with Maxwell's equations for the fields), an appropriate expression for P_a, i.e. a proper equation of state (EoS) must be chosen. As an example, for electron plasma waves in a warm plasma $T_e n_e^{-\gamma_e + 1} = P_e n_e^{-\gamma_e} = $ const. holds with the adiabatic coefficient $\gamma_e = 3$ (see e.g. Mulser and Bauer 2010, Sect. 2.2.2), where T_e is the electron temperature (energy units for which the Boltzmann constant $k_B = 1$ are assumed).

As a case which will be of interest in Chap. 5, let us consider an electrostatic regime with no magnetic field, so that $\mathbf{E} = -\nabla \Phi$, and where we are interested on phenomena occurring on the time scale of ion motion. Thus we assume that on such scale the electrons rapidly respond to changes in the field and assume a mechanical equilibrium condition with $\mathbf{u}_e = 0$ (or, equivalenty, we assume the electrons to have zero mass), so that (2.71) yields $e n_e \nabla \Phi - \nabla P_e = 0$. We further assume that the electrons have an isothermal, perfect gas-like EoS $P_e = n_e T_e$ with constant and uniform temperature T_e. By integrating the resulting equation $(\nabla n_e)/n_e = e \nabla \Phi / T_e$ we obtain the Boltzmann equilibrium condition

$$n_e = n_0 \exp(e\Phi/T_e), \qquad (2.72)$$

which is probably already known by undergraduate readers who met Debye screening as an example of elementary plasma physics (see e.g. Feynman et al. 1963, Sect. 7.4). Equation (2.72) with Poisson's equation

$$\nabla^2 \Phi = -4\pi \varrho = -4\pi e(n_i - n_e), \qquad (2.73)$$

and Eqs. (2.67, 2.71) for ions yield a consistent model which, when the equations are linearized, provides the dispersion relation for ion-acoustic waves. The model, keeping the nonlinear terms, will be used to describe the expansion of a warm plasma (Sect. 5.3) and the formation of collisionless shock waves (Sect. 5.5), both phenomena being of current interest for laser-plasma acceleration of ions.

The other choice for the EoS which will be considered in the following is indeed a trivial one, $P_a = 0$. This choice actually corresponds to assuming $|\mathbf{w}| \ll v_a$: the thermal velocity is negligible with respect to the average, coherent velocity. In very intense laser-plasma interactions the electron motion is typically dominated by the coherent oscillation in the EM fields, so that the above assumption is a reasonable one. Neglecting P_a brings to the so-called *cold* fluid equations, although this definition may sound strange when referring to plasmas whose energy density is larger by order of magnitudes than for genuine low-temperature plasmas as found in electrical discharges (not to mention *ultracold* plasmas created by photoionization and having temperatures lower than 1 mK). When a negligible thermal spread is assumed *a priori* in the derivation of fluid equations (so that effectively the momentum distribution is a Dirac delta), the relativistic version of the latter is obtained quite easily by naturally defining the average momentum $\mathbf{p}_a = \mathbf{p}_a(\mathbf{r}, t) = \int \mathbf{p} f_a d^3 p$. The relativistic, "cold fluid" momentum equation becomes

$$(\partial_t \mathbf{p}_a + \mathbf{u}_a \cdot \nabla \mathbf{p}_a) = q_a \left(\mathbf{E} + \mathbf{u}_a \times \mathbf{B}/c \right), \qquad \mathbf{p}_a = m_a \gamma_a \mathbf{u}_a. \qquad (2.74)$$

In contrast, when thermal effects are not neglected the choice of the EoS for the closure of the fluid equations and the derivation of the correct form of the latter are definitely not trivial issues (see e.g. Pegoraro and Porcelli 1984).

Notice that in Eq. (2.74) neglecting P_a allowed to cancel out the density n_a. If we substitute d/dt for $(\partial_t + \mathbf{u}_a \cdot \nabla)$, Eq. (2.74) now looks identical to the equation of motion (2.13) for the single particle. To avoid possible confusion it might be worth to stress that in Eq. (2.74) $\mathbf{p}_a$ and $\mathbf{u}_a$ are Eulerian variables which depend on $(\mathbf{r}, t)$ coordinates as the EM fields, while in Eq. (2.13) $\mathbf{p} = \mathbf{p}(t)$ and $\mathbf{r} = \mathbf{r}(t)$ are functions of time describing the particle trajectory and the EM fields in the Lorentz force need to be evaluated along the trajectory itself, i.e. $\mathbf{E} = \mathbf{E}(\mathbf{r}, t)$ and $\mathbf{B} = \mathbf{B}(\mathbf{r}, t)$. As an example which may highlight this fundamental difference, one may derive a ponderomotive force density via the fluid equations using an iterative procedure analogous to that presented in Sect. 2.1.4; in doing so, one realizes that a nonlinear contribution now comes from the $\mathbf{u}_a \cdot \nabla \mathbf{p}_a$ term while a Taylor expansion of the fields along the trajectory makes obviously no sense.

2.3 Simulation Models

Both in the kinetic and in the fluid description, the equations describing a collisionless plasma in self-consistent electromagnetic fields are strongly nonlinear. Analytical solutions are hard to find and in a limited number. Thus, a crucial role is played by simulations, i.e. by the numerical solution of the model equations for given geometry and boundary conditions. The continuous increase in computing power allows simulations to consider progressively larger system sizes and calculation times approaching those typical of experimental observations. At the same time, relatively complex simulations may be already performed on cheap personal computers. For these reasons, a larger part of the laser-plasma community is now using simulation codes, some of the latter being freely available.

Nowadays numerical simulations are essential tools of physics, particularly of plasma physics, and their importance and utility in the study of laser plasmas may hardly be overestimated. Still, it is important that the users do not always use a code written by someone else as a black box, but have some insight about what is inside the code, including a basic knowledge of algorithms and numerical methods. Of course, the most sophisticated codes require years of development and continuous upgrading, so that teamwork and inheritance of other contributions will be essential for those newcomers interested in high-performance computing. However, in our opinion it is very useful that students and novices in computational laser-plasma physics write and run a code (even a simple one) starting from scratch at least once in a lifetime.

As a suggestion for a model adequate as a primer for plasma simulations, in the next Sect. 2.3.1 we introduce the "sheet" model by Dawson (1962). Published a few years after the celebrated work by Fermi, Pasta, Ulam and Tsingou (see Dauxois 2008) which is considered to be the first-ever numerical experiment in physics, Dawson's paper (Dawson 1962) may be regarded as the beginning of plasma simulation research. Although the model is purely electrostatic, it is still useful to illustrate phenomena of interest for laser-plasma physics such as plasma wake generation (Sect. 4.1) and collisionless heating (Sect. 4.2.2), which will be discussed in the following. Then, in Sect. 2.3.2 we describe the basics of the Particle-In-Cell (PIC) method, which is the mostly used approach to laser-plasma numerical modeling.

2.3.1 Dawson's Sheet Model

In Dawson's model an one-dimensional planar geometry is assumed with electrons being effectively charged *sheets* moving along the x axis. Ions are taken as an immobile neutralizing background. Let $i = 1, 2, \ldots N$ be the index of the N sheets composing the system. In absence of external forces, at equilibrium the sheets are in the positions $x = X_i^0$ equally spaced along x and, following Gauss's law, the electric field has a sawtooth shape and its average on each sheet vanishes (see Fig. 2.4). When a sheet is displaced out of the equilibrium position, assuming for the moment that the sheet does not overturn the neighboring sheets the average field at the sheet position X_i becomes proportional to the displacement (still according to Gauss's law), $E_x(X_i) = 4\pi e n_0 (X_i - X_i^0)$ with n_0 the background density. The equation of motion for the position $X_i = X_i(t)$ of each sheet may thus be written as

$$\frac{d^2 X_i}{dt^2} = -\frac{e}{m_e} E_x(X_i) + \frac{1}{m_e} F_{\text{ext}} = -\omega_p^2 (X_i - X_i^0) + \frac{1}{m_e} F_{\text{ext}}, \qquad (2.75)$$

where F_{ext} represents any external force. The above harmonic oscillator equation indicates that any "small" perturbation (which does not make the sheets to "collide") drives oscillations at the plasma frequency $\omega_p = (4\pi e^2 n_0 / m_e)^{1/2}$. Now consider the

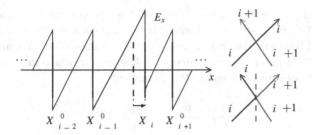

Fig. 2.4 *Left frame* sketch of the spatial distribution of electron "sheets" and of the electric field in Dawson's model. Only the ith sheet is shown out of its equilibrium position. *Right frame* schematic showing how the crossing of neighboring sheets is implemented as an elastic collision, equivalent to an exchange of the sheet indices

case when, at a time t, the sheet with index i gets over the position of the neighboring one (say, $i + 1$). The mutual crossing of the two sheets cause a jump in the amount of charge on each side of the sheets and correspondingly in the electric field. To make the computation easy, the indices of the two sheets are exchanged, i.e. $i \leftrightarrow i + 1$, which is equivalent to a "remapping" of initial conditions at time t (see Fig. 2.4). In this way, it is possible to keep the equation of motion (2.75). In practice, the exchange of indices corresponds to the exchange of velocities as if the two sheets underwent an elastic collision. Nonlinearity is introduced in the model by such artificial collision, which is easily implemented at each time step in the simulation. The integration of (2.75) may be performed by any standard numerical method for ordinary differential equations.

In the original paper Dawson (1962) a statistical mechanics analysis of the model is discussed. When an initial "temperature" distribution is given to the sheets, the system exhibits Debye screening and Landau damping, which are concepts familiar to plasma physics students. As another model problem, oriented to the study of stopping of fast charged particles in a plasma, one of the sheets is given a large initial velocity and a *wake* of plasma oscillations is excited as the sheet propagates and is decelerated by the electric field associated to the wake. As will be discussed in Sect. 4.1 the generation of wakefields is the basis of laser-plasma acceleration of electrons. Their elementary features may be investigated with the sheet model.

2.3.2 Particle-In-Cell Method

The PIC method to solve the kinetic equation (2.63) numerically is based on the following representation for the distribution function

$$f(\mathbf{r}, \mathbf{p}, t) = A \sum_{l=0}^{N_p - 1} g[\mathbf{r} - \mathbf{r}_l(t)]\delta^3[\mathbf{p} - \mathbf{p}_l(t)], \qquad (2.76)$$

where $\mathbf{r}_l(t)$ and $\mathbf{p}_l(t)$ are functions of time and A is a proper normalization constant. In (2.76) $\delta^3(\mathbf{p}) = \delta(p_x)\delta(p_y)\delta(p_z)$ where $\delta(p)$ is the Dirac function, and $g(\mathbf{r})$ is an even, localized function (with properties similar to δ^3) that will be specified later. By substituting (2.76) into (2.63) and integrating first over d^3p and then over d^3r, the following equations for $\mathbf{r}_l(t)$ and $\mathbf{p}_l(t)$ are obtained:

$$\frac{d\mathbf{r}_l}{dt} = \frac{\mathbf{p}_l}{m_a\gamma_l}, \qquad \frac{d\mathbf{p}_l}{dt} = \bar{\mathbf{F}}_l, \tag{2.77}$$

where

$$\bar{\mathbf{F}}_l = \bar{\mathbf{F}}_l(\mathbf{r}_l, \mathbf{p}_l, t) = \int g[\mathbf{r} - \mathbf{r}_l(t)]\mathbf{F}(\mathbf{r}, \mathbf{p}, t)d^3r. \tag{2.78}$$

The problem is thus reduced to $2N_p$ equations of motion (2.77), describing the motion of N_p computational particles. This sounds like going back to the fundamental description of a plasma as an ensemble of charged particles, but there are two fundamental differences with such an "ab initio" approach. First, the number of computational particles which may be allocated on a computer will be almost always lower by several order of magnitudes than the actual number of electrons and ions in the system under investigation, so that we actually have a very limited sample of the phase space. Second, the particles are accelerated by the mean EM fields, consistently with Vlasov–Maxwell theory, and interparticle interactions are not computed.

The particles are point-like in momentum space (i.e. they have $\mathbf{p}_l$ as a single definite value of the momentum) but are in general "extended" in coordinate space. From Eq. (2.78) we infer that $g(\mathbf{r})$ describes the "shape", i.e. the profile of the charge distribution associated to a particle, as it is further apparent by computing the charge and current densities from (2.76) according to (2.64, 2.65):

$$\varrho(\mathbf{r}, t) = A \sum_{a,l} q_a g[\mathbf{r} - \mathbf{r}_{a,l}(t)], \qquad \mathbf{J}(\mathbf{r}, t) = A \sum_{a,l} q_a \mathbf{v}_l g[\mathbf{r} - \mathbf{r}_{a,l}(t)]. \tag{2.79}$$

In the PIC approach, ϱ and $\mathbf{J}$ as well as the EM fields are discretized on a spatial grid. The position of the particle with respect to the grid and its charge distribution described by $g(\mathbf{r})$ determine both the contribution to the total current in the grid cells overlapping with the particle and the average force on the particle, according to Eq. (2.78) that also reduces to a sum over overlapping cells. The weights in the sum are proportional to the volume of the overlapping regions which, for a given $g(\mathbf{r})$, are well defined analytical functions of the particle and cell positions. Figure 2.5 shows the typical case of a triangular shape for which the particle overlaps to 3^D cells where D is the dimensionality of the problem. Although in principle also $g(\mathbf{r})$ may be a delta function describing a point particle, this choice is rarely used since using "cloud-like" particles smooths out the resulting distribution and reduces noise fluctuations (crossing of cell boundaries by a point particle leads to an abrupt jump of the density at the cell location).

Fig. 2.5 Example showing
a particle moving across the
grid in the PIC approach.
The particle is described
by a "triangular" density
profile $g(x)$ (normalized to
unity) whose width is twice
the grid spacing Δx. The
particle thus overlaps to three
cells in 1D and to nine cells
in 2D (see *upper frame*),
where its shape function is
$g(\mathbf{r}-\mathbf{r}_l) = g(x-x_l)g(y-y_l)$

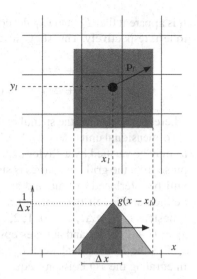

The PIC method is extensively described in the classic book by
Birdsall and Langdon (1991). The book contains a description of the most widely
used methods for the essential parts of a PIC code, such as the leapfrog algorithm
to advance particle positions, the Boris pusher scheme to advance momenta, explicit
finite difference algorithms to integrate Maxwell's equations, and formulas for shape
factors to compute both the charge and current densities on the grid and the average
force on the particle, see Eqs. (2.78) and (2.79). Hereby we summarize such information in the 1D case, in order to give a quick PIC primer (see also Gibbon 2005,
2012). Details on basic numerical concepts, such as the discretized form of differential operators, may be found in introductory books on scientific computing, see e.g.
Press et al. (2007).

For codes developed for laser-plasma physics it is convenient to use the following
normalizations: time in units of ω^{-1} with ω the laser frequency, space in units of
c/ω, momenta in units of $m_e c$, fields in units of $m_e \omega c / e$, and densities in units
of $n_c = m_e \omega^2 / 4\pi e^2$. In 1D geometry with x as the only spatial coordinate, the
normalized Maxwell's equations for the transverse fields are

$$\partial_t E_y = -J_y - \partial_x B_z, \qquad \partial_t B_z = -\partial_x E_y, \qquad (2.80)$$
$$\partial_t E_z = -J_z + \partial_x B_y, \qquad \partial_t B_y = \partial_x E_z, \qquad (2.81)$$

which show that the two possible states of linear polarization are uncoupled. To
integrate these equations, we introduce the auxiliary fields $F_\pm \equiv E_y \pm B_z$ and
$G_\pm \equiv E_z \pm B_y$ which satisfy the equations

$$(\partial_t \pm \partial_x)F_\pm = -J_y, \qquad (\partial_t \pm \partial_x)G_\mp = -J_z. \qquad (2.82)$$

It is apparent that $F_\pm$ and $G_\mp$ describe waves propagating from left to right and right to left, respectively. This suggests the following scheme to integrate (2.82)

$$F_\pm(x \pm \Delta x, t + \Delta t) = F_\pm(x, t) - J_y(x \pm \Delta x/2, t + \Delta t/2)\Delta t, \qquad (2.83)$$
$$G_\pm(x \mp \Delta x, t + \Delta t) = G_\pm(x, t) - J_z(x \mp \Delta x/2, t + \Delta t/2)\Delta t, \qquad (2.84)$$

where $\Delta x = \Delta t$, i.e. the spatial resolution (cell width) is taken equal to the timestep (in dimensional units, $\Delta x = c\Delta t$). The scheme is second-order accurate ($\sim\mathcal{O}(\Delta t^2)$), as may be verified by a Taylor expansion. With such scheme, the injection of a laser pulse from the grid boundaries is straightforward. Notice that the transverse currents will be interlaced in time and space with the transverse fields: if $F_\pm$ and $G_\pm$ are defined at the cell boundaries ($x_i = i\Delta x, i = 0, 1, 2, \ldots, N - 1$) and at integer timesteps $t_n = n\Delta t, n = 0, 1, 2, \ldots$, then J_y and J_z are defined at the cell centers $x_i = (i + 1/2)\Delta x$ and at timesteps $t_n = (n + 1/2)\Delta t$.

For the longitudinal field E_x, a quick explicit scheme is obtained directly by integrating the 1D Poisson's equation $\partial_x E_x = \varrho$:

$$\begin{aligned} E_x(x, t) &= E(x - \Delta x, t) + \int_{x-\Delta x}^{x} \varrho(x', t)dx' \\ &= E(x - \Delta x, t) + \varrho(x - \Delta x/2, t)\Delta x + \mathcal{O}(\Delta x^2). \end{aligned} \qquad (2.85)$$

The charge density must therefore be defined at cell centers. Alternatively, one can use an implicit method to obtain E_x via the electrostatic potential Φ, from the equation $\partial_x^2 \Phi = -\varrho$ which in discretized form yields

$$\frac{\Phi(x - \Delta x) - 2\Phi(x) + \Phi(x + \Delta x)}{\Delta x^2} + \mathcal{O}(\Delta x^2) = -\varrho(x). \qquad (2.86)$$

In grid coordinates $x_i = i\Delta x$ with $i = 0, 1, \ldots, N-1$ (2.86) becomes a linear system of N equations whose matrix of coefficients is tridiagonal, and may be solved by standard numerical methods of matrix inversion.

To compute the charge and current densities in each cell according to (2.79), as stated above the contribution of each particle to a cell is proportional to the volume of the overlapping region. For the 1D case and the triangular shape of Fig. 2.5, each particle contributes to three cells, with weights given by the area of the shaded regions in the bottom frame of Fig. 2.5, divided by the cell length Δx. The weighting factors are a function of the relative position $x_l - x_i$, where x_l and x_i are the positions of the particle and of the center of the parent cell (such that $|x_l - x_i| < \Delta x/2$), respectively. Explicitly, for the triangular shape

$$S_{-1} = \frac{1}{2}\left(u - \frac{1}{2}\right)^2, \quad S_0 = \left(\frac{3}{4} - u^2\right), \quad S_{+1} = \frac{1}{2}\left(u + \frac{1}{2}\right)^2, \qquad (2.87)$$

where $u = (x_l - x_i)/\Delta x$. According to (2.78), the same weights must be also used in the calculation of the average force on each particle using the values of the fields in the overlapping cells. Thus, for example, if i is the index of the parent cell, the average electric field on the particle is

$$\bar{E}_l = \sum_{j=-1}^{j=+1} S_j(u)E_{i+j},\qquad(2.88)$$

where E_j is the field in the $(i+j)$th cell, with $i+j = i-1, i, i+1$. (Some attention has to be paid when the fields are defined at different positions, e.g. at the cell edges, with respect to the densities).

To advance the particle momenta in time, we describe the widely used algorithm known as the Boris pusher (notice that the scheme operates in a 3D momentum space and may thus be used also in 2D and 3D codes). We start from the knowledge of the particle momentum $\mathbf{p}$ at the timestep $n - 1/2$ and of the EM fields (averaged on the particle) at the timestep n (we drop the particle index for simplicity). First we make an "half-boost" by the electric field:

$$\mathbf{p}^{(-)} = \mathbf{p}^{(n-1/2)} + \frac{q}{m}\bar{\mathbf{E}}^{(n)}\frac{\Delta t}{2}.\qquad(2.89)$$

Then, the momentum is rotated in the magnetic field. Defining

$$\gamma^{(n)} = \sqrt{1 + (\mathbf{p}^{(-)}/mc)^2}, \quad \mathbf{t} \equiv \frac{q\bar{\mathbf{B}}^n}{m\gamma^n c}, \quad \mathbf{s} \equiv \frac{2\mathbf{t}}{1+t^2},\qquad(2.90)$$

the rotation occurs in two steps:

$$\mathbf{p}' = \mathbf{p}^{(-)} + \mathbf{p}^{(-)} \times \mathbf{t}, \quad \mathbf{p}^{(+)} = \mathbf{p}^{(-)} + \mathbf{p}' \times \mathbf{s}.\qquad(2.91)$$

Finally, another half-boost by the electric field is performed to complete the momentum update:

$$\mathbf{p}^{(n+1/2)} = \mathbf{p}^{(+)} + \frac{q}{m}\bar{\mathbf{E}}^{(n)}\frac{\Delta t}{2}.\qquad(2.92)$$

To advance particle positions, use is made of a leapfrog scheme, that requires the coordinate to be defined at integer timesteps:

$$\hat{\mathbf{x}}^{(n+1)} = \hat{\mathbf{x}}^{(n)} + \mathbf{v}^{(n+1/2)}\Delta t, \quad \mathbf{v}^{(n+1/2)} = \frac{\mathbf{p}^{(n+1/2)}}{m\gamma^{(n)}}.\qquad(2.93)$$

Both the Boris pusher and the leapfrog scheme are second order accurate.

PIC codes in 2D and 3D usually employ the finite-difference time-domain (FDTD) technique to compute the EM fields, which are distributed on the grid according to a scheme known as the Yee lattice (Yee 1966). Notice that the fields are advanced using the equations $\partial_t \mathbf{B} = -c\nabla \times \mathbf{E}$ and $\partial_t \mathbf{E} = c\nabla \times \mathbf{B} - 4\pi \mathbf{J}$ only, and the scheme preserves the condition $\nabla \cdot \mathbf{B} = 0$ if initially imposed. However, depending on the field and current distribution on the grid, it is possible that the equation $\nabla \cdot \mathbf{E} = 4\pi \varrho$ (which is not used to calculate the fields) is not fulfilled exactly. This circumstance may be prevented by an advanced scheme for the reconstruction of $\mathbf{J}$ from the particle distribution that ensures the continuity equation $\partial_t \varrho = -\nabla \cdot \mathbf{J}$ to hold exactly. Esirkepov (2001) describes such a scheme with the aim to make simulations of laser interaction with overdense plasmas more accurate.

A few general suggestions on the set-up of PIC simulations might be worth giving. The problem to be investigated via simulations will typically have a smallest scale ℓ which needs to be resolved, i.e. the spatial resolution (or cell size) Δx must be quite smaller that ℓ. For laser-plasma interactions, typically $\ell = \lambda$ (the laser wavelength) for underdense plasmas and $\ell = c/\omega_p$ (the collisionless skin depth) for overdense plasmas (see Sect. 3.1). Usually the temporal resolution is bounded by numerical stability constraints, such as $\Delta t \leq \Delta x/c$ in 1D. To estimate the number of particles per cell (N_p) to be allocated initially, let us imagine the density to be initially constant and equal to n_0. Then, an isolated particle would yield in the overlapping cells a density $\sim n_0/N_p$, which may be then considered the smallest density value that can be resolved accurately. A large value of N_p is then needed to resolve large density gradients and describe low density regions. Using many particles is also essential to reduce statistical noise and to take into account tails of very energetic particles, which may be in small numbers but either play an essential role in the dynamics or be the most interesting quantity to observe. On the other hand N_p will be limited because of the available memory. Assuming for simplicity that the spatial grid has a size $\simeq L$ and a resolution $\simeq \Delta x$ along all the axes, the number of gridpoints is $N_g = (L/\Delta x)^D$ where D is the dimensionality of the problem, and the total number of particles scales as $N_p N_g f$ where f is the fraction of the grid initially occupied by the plasma. Each particle is typically represented by $D + 3$ numbers for spatial and momentum coordinates, each requiring 8 bytes, so that eventually the total memory may be estimated as $8(D+3)N_p(L/\Delta x)^D f$ bytes assuming that the load due to the particles dominates over that due to the EM fields (which is $\sim 8 \times 6 \times N_g$). It will be then apparent that simulating high density plasmas for $D > 1$ is a very demanding task, so that both the use of supercomputing (which requires a parallel PIC code) and the search of a reasonable compromise between "real" and simulations parameters (such as lower density, smaller scales, ...) will be necessary.

2.3.3 Boosted Frame Simulations

The considerations on the size of "realistic" simulations of the preceding section should make apparent that any strategy able to reduce the computational cost is welcome. Relativity provides an help on this matter, as a change of reference frame, i.e. a Lorentz boost, may be effective to strongly reduce the simulation size in suitable situations.

The first example we discuss concerns the problem of the oblique incidence at an angle θ of a plane wave pulse on a planar surface in the laboratory frame S, as sketched in Fig. 2.6. This is a two-dimensional problem, where all the fields and currents depend on the (x, y) coordinates. However, a Lorentz boost in the direction parallel to the surface with velocity parameter $\beta = \sin \theta$ turns oblique incidence into *normal* incidence in the *boosted* frame S', where all quantities are now independent on y'. In fact, the Lorentz transformations of the frequency ω and the wavevector $\mathbf{k} = (\omega/c)(\cos \theta, \sin \theta)$ give

$$k'_x = k_x = k \cos \theta = k/\gamma, \tag{2.94}$$

$$k'_y = \gamma(k_y - (\omega/c)\beta) = \gamma(\omega/c)(\sin \theta - \sin \theta) = 0, \tag{2.95}$$

$$\omega' = \gamma(\omega - k_y \beta) = \gamma\omega(1 - \sin^2 \theta) = \omega/\gamma. \tag{2.96}$$

Thus, $k'_y = 0$ in S' and the incidence is normal. The frequency $\omega' = k'c$ is decreased by the factor $\gamma = 1/\cos \theta$. The phase $k_y y - \omega t$ common to all fields and currents in S transforms according to $k_y y - \omega t = k'_y y' - \omega' t' = -\omega' t'$, so that in S' only the dependence on the $x' = x$ coordinate remains.

For a P-polarized wave, as in Fig. 2.6, the transformation of the EM fields yield

$$E'_x = \gamma(E_x + \beta B_z) = \gamma E(\sin \theta - \sin \theta) = 0, \tag{2.97}$$

$$E'_y = E_y = E \cos \theta = E/\gamma, \tag{2.98}$$

$$B'_z = \gamma(B_z + \beta E_x) = \gamma B(1 - \sin^2 \theta) = B/\gamma. \tag{2.99}$$

In S' there is not an electric field component perpendicular to the surface, as it is obvious for a transverse wave at normal incidence. One may thus wonder if the well-known difference in the reflection and transmission of EM waves between S- and P-polarizations, that is apparent in Fresnel formulas, is lost in S'. The point is that one may still distinguish between S and P even at normal incidence because the isotropy of the medium is broken by its drift at the velocity $\mathbf{V} = -\beta c \hat{\mathbf{y}}$ in S'. This leads to the appearance of a first-order force on the charges in the medium at frequency ω and perpendicular to the surface for P-polarization (that may now be defined as the case of electric field parallel to $\mathbf{V}$): the Lorentz force in the x direction is $F'_x = qv'_y B'_z$, i.e. $qc \sin \theta B'$ to lowest order. In contrast, there is no such coupling for S-polarization in which $\mathbf{B}'$ is parallel to $\mathbf{V}$. As it will be discussed in Sect. 4.2, the longitudinal force components play a crucial role in the laser-plasma coupling at the surface of solid targets.

The boosted frame transformation thus turns a 2D problem into a 1D one. Originally proposed as a trick to simplify the calculations of resonance absorption (Bourdier 1983) (see Sect. 4.2.1), the transformation has been later implemented into simulation codes (Gibbon and Bell 1992) yielding a save of computational costs by a factor equal to the number of gridpoints one would need for the additional dimension

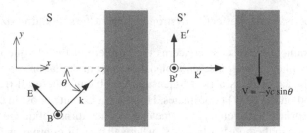

Fig. 2.6 A plane EM wave obliquely incident at an angle θ in the S frame becomes at normal incidence in the *boosted* frame S', with the transformation velocity $\boldsymbol{\beta} = \hat{\mathbf{y}} \sin \theta$. The medium drifts with velocity $\mathbf{V} = -c\boldsymbol{\beta}$ in S'. The EM fields are shown for P-polarization

in the 2D case. As both Maxwell's equations and the relativistic equations of motion are covariant, in any 1D PIC code the use of the technique substantially corresponds in a proper initialization and transformations of the laser and plasma parameters in the boosted frame. One should take care, however, to transform the parameters correctly, especially since in a code all quantities are typically normalized to parameters which may be not Lorentz invariants (see Gibbon et al. (1999) for a discussion). In addition, one should be aware that the physics may look different in the boosted frame: for example, electrostatic waves acquire a magnetic component because of the Lorentz transformations, a surface charge in S leads to a surface current in S' because of the drift velocity, and so on.

More recently, it has been suggested that a Lorentz boost transformation allows a substantial computational saving for a different class of problems (Vay 2007) which correspond, in the present context, to the propagation of a laser pulse of duration τ_L through a plasma of length L_P, a set-up of direct interest for laser-plasma acceleration of electrons (see Sect. 4.1). The argument of Vay (2007) is outlined as follows. In general, the number of gridpoints along a given direction in a simulation scales with the ratio between the spatial extension of the system that we aim to simulate ("long" scale) and the smallest length that need to be resolved for physical accuracy ("small" scale). Similarly, the number of timesteps scales with the ratio of the time we aim to simulate and the shortest time to be resolved. In the present problem, one has at least to resolve the laser wavelength λ and period $T = \lambda/c = 2\pi/\omega$, so we assume these latter as the short scales. The long spatial scale is the sum of the plasma and the pulse length, $L_p + c\tau_L$. The long temporal scale is the "interaction time" $\tau_I \simeq L_p/c + \tau_L$, i.e. the time needed for the pulse to cross the plasma. The ratio between the long and the short scales is thus $R = (L_p + c\tau_L)/\lambda$ for both space and time. Now consider a boosted frame moving with velocity βc parallel to the pulse propagation. The length of the plasma is contracted down to $L'_p = L_p/\gamma$, while the laser wavelength, period and frequency transform according to $\lambda' = cT' = \gamma(1 + \beta)\lambda = \gamma(1 + \beta)cT$ and $\omega' = \omega/(\gamma(1+\beta))$, so that the pulse duration $\tau'_L = \gamma(1+\beta)\tau_L$. The interaction time $\tau'_I = (L'_p + c\tau'_L)/(c(1 + \beta))$ in the boosted frame, because of the drift of the plasma at the velocity $-\beta c$. Thus, the ratio between the largest and the smallest scales is now different for space and time scales:

$$R'_t = \frac{\tau'_I}{T'} = \frac{L_p/\lambda}{\gamma^2(1+\beta)^2} + \frac{c\tau_L/\lambda}{1+\beta}, \tag{2.100}$$

$$R'_s = \frac{L'_p + c\tau'_L}{\lambda'} = \frac{L_p/\lambda}{\gamma^2(1+\beta)} + \frac{c\tau_L}{\lambda}. \tag{2.101}$$

These expressions suggest that both R'_t and R'_s may be strongly reduced in S': the effect is large on the L_p/λ term, which is usually much larger than $c\tau_L/\lambda$ for simulations of laser wakefield acceleration. Transforming in the boosted frame, the number of spatial grid cells actually remains the same because the number of cycles within the laser pulse is invariant, but the contraction of L_p and the increase of λ (and thus of the cell length) leads to a shorter interaction length and allows for a larger timestep. Thus, the use of the boosted frame allows to reduce the computational time by a factor of the order of $R/R'_t \simeq \gamma^2(1+\beta^2)$, that is a very large number as one follows the propagation of the laser pulse at a velocity $v_g \lesssim c$, so $\gamma^2 \gg 1$ if $\beta = v_g$. Such computational gain has allowed to extend laser wakefield acceleration over very large scales (Martins et al. 2010), as required by the foreseen developments of laser-plasma accelerators.

References

Birdsall, C.K., Langdon, A.B.: Plasma Physics Via Computer Simulation. Institute of Physics, Bristol (1991)
Bourdier, A.: Phys. Fluids **26**, 1804 (1983)
Capdessus, R., d'Humières, E., Tikhonchuk, V.T.: Phys. Rev. E **86**, 036401 (2012)
Chen, M., Pukhov, A., Yu, T.P., Sheng, Z.M.: Plasma Phys. Control Fusion **53**, 014004 (2011)
Dauxois, T.: Phys. Today **61**, 55–57 (2008). (January issue)
Dawson, J.: Phys. Fluids **5**, 445 (1962)
Dawson, J.M.: Rev. Mod. Phys. **55**, 403 (1983)
Di Piazza, A.: Lett. Math. Phys. **83**, 305 (2008)
Di Piazza, A., Hatsagortsyan, K.Z., Keitel, C.H.: Phys. Rev. Lett. **102**, 254802 (2009)
Di Piazza, A., Hatsagortsyan, K.Z., Keitel, C.H.: Phys. Rev. Lett. **105**, 220403 (2010)
Esirkepov, T.Z.: Comput. Phys. Commun. **135**, 144 (2001)
Feynman, R.P., Leighton, R.B., Sands, M.: The Feynman Lectures on Physics, vol. 2. Addison-Wesley, Reading (1963)
Gibbon, P.: Short Pulse Laser Interaction with Matter. Imperial College Press, London (2005)
Gibbon, P.: Rivista del Nuovo Cimento **35**, 607 (2012)
Gibbon, P., Bell, A.R.: Phys. Rev. Lett. **68**, 1535 (1992)
Gibbon, P., et al.: Phys. Plasmas **6**, 947 (1999)
Haberberger, D., Tochitsky, S., Joshi, C.: Opt. Express **18**, 17865 (2010)
Jackson, J.D.: Classical Electrodynamics, 3rd edn. Wiley, New York (1998)
Kim, K.J., McDonald, K.T., Stupakov, G.V., Zolotorev, M.S.: Phys. Rev. Lett. **84**, 3210 (2000)
Landau, L.D., Lifshitz, E.M.: The Classical Theory of Fields, 2nd edn. Elsevier, Oxford (1975)
Martins, S.F., Fonseca, R.A., Lu, W., Mori, W.B., Silva, L.O.: Nat. Phys. **6**, 311 (2010)
McDonald, K.T.: Phys. Rev. Lett. **80**, 1350 (1998)
Mora, P., Quesnel, B.: Phys. Rev. Lett. **80**, 1351 (1998)
Mulser, P., Bauer, D.: High Power Laser-Matter Interaction. Springer, Berlin (2010)
Naumova, N., et al.: Phys. Rev. Lett. **102**, 025002 (2009)

Pegoraro, F., Porcelli, F.: Phys. Fluids **27**, 1665 (1984)

Peierls, R.E.: Surprises in Theoretical Physics. Princeton University Press, Princeton (1979)

Press, W.H., Teukolsky, S.A., Vetterling, W.T., Flannery, B.P.: Numerical Recipes Third Edition:
   The Art of Scientific Computing, 3rd edn. Cambridge University Press, New York (2007)

Tamburini, M., Pegoraro, F., Di Piazza, A., Keitel, C.H., Macchi, A.: New J. Phys. **12**, 123005
   (2010)

Troha, A.L., Hartemann, F.V.: Phys. Rev. E **65**, 028502 (2002)

Vay, J.L.: Phys. Rev. Lett. **98**, 130405 (2007)

Yanovsky, V., et al.: Opt. Express **16**, 2109 (2008)

Yee, K.: IEEE Trans. Antennas Propag. **14**, 302 (1966)

Chapter 3
Relativistic Nonlinear Waves in Plasmas

Abstract In this chapter we focus on waves in a relativistic plasma. For electromagnetic waves, we introduce the nonlinear refractive index and the two most prominent phenomena of "relativistic optics", i.e. self-focusing and transparency. For both phenomena, an account of a more complete theoretical description is presented along with an introduction to some methods of nonlinear physics, such as the multiple scale expansion, the nonlinear Schrödinger equation, and the Lagrangian approach. A brief description of standing nonlinear solutions, i.e. cavitons or (post-)solitons, is also given. For electrostatic waves we discuss the wave-breaking limit and focus on properties relevant to electron accelerators and field amplification schemes that will be described in the following chapters.

3.1 Linear Waves

The starting point of our analysis is the wave equation for the electric field

$$\left(\nabla^2 - \frac{1}{c^2}\partial_t^2\right)\mathbf{E} - \nabla(\nabla \cdot \mathbf{E}) = \frac{4\pi}{c^2}\partial_t \mathbf{J}, \tag{3.1}$$

which is obtained from Maxwell's equation by eliminating $\mathbf{B}$.

Here we consider an unmagnetized, "cold" plasma whose response to high-frequency fields is due to electrons only. The current density $\mathbf{J} = -en_e\mathbf{u}_e$ may thus be obtained by the cold fluid equations of Sect. 2.2.2.

First we review *linear* waves in a homogeneous plasma with uniform and constant electron density n_e, neglecting all nonlinear and "relativistic" terms. Thus we linearize Eq. (2.71) obtaining $\partial_t \mathbf{J} \simeq -(e/m_e)n_e\mathbf{E}$. For monochromatic fields, using the same notation as in Eq. (2.38) we obtain

$$\tilde{\mathbf{J}} = -i\frac{n_0 e^2}{m_e \omega}\tilde{\mathbf{E}} = -\frac{i}{4\pi}\frac{\omega_p^2}{\omega}\tilde{\mathbf{E}}, \tag{3.2}$$

A. Macchi, *A Superintense Laser-Plasma Interaction Theory Primer*,
SpringerBriefs in Physics, DOI: 10.1007/978-94-007-6125-4_3,
© The Author(s) 2013

with the plasma frequency

$$\omega_p \equiv \left(\frac{4\pi e^2 n_e}{m_e}\right)^{1/2}. \tag{3.3}$$

By substituting into (3.1) we obtain the inhomogeneous Helmholtz equation

$$\left(\nabla^2 + \varepsilon(\omega)\frac{\omega^2}{c^2}\right)\tilde{\mathbf{E}} - \nabla(\nabla \cdot \tilde{\mathbf{E}}) = \left(\nabla^2 + \mathsf{n}^2(\omega)\frac{\omega^2}{c^2}\right)\tilde{\mathbf{E}} - \nabla(\nabla \cdot \tilde{\mathbf{E}}) = 0, \tag{3.4}$$

where we introduced the well-known expressions for the dielectric function $\varepsilon(\omega)$ and the refraction index $\mathsf{n}(\omega)$ of a cold plasma

$$\varepsilon(\omega) = \mathsf{n}^2(\omega) = 1 - \frac{\omega_p^2}{\omega^2}. \tag{3.5}$$

Now consider transverse EM waves with $\nabla \cdot \mathbf{E} = 0$. For a plane wave $\tilde{\mathbf{E}}(\mathbf{r}) = \mathbf{E}_0 e^{i\mathbf{k}\cdot\mathbf{r}}$ with $\mathbf{k} \cdot \mathbf{E}_0 = 0$, we obtain by direct substitution the dispersion relation

$$-k^2 c^2 + \varepsilon(\omega)\omega^2 = -k^2 c^2 + \omega^2 - \omega_p^2 = 0. \tag{3.6}$$

The propagation of the wave requires $k = |\mathbf{k}|$ to be a real number, which occurs when $\omega > \omega_p$. Thus, the plasma frequency is a cut-off frequency for EM transverse waves. For a given frequency ω this condition can be also written as a condition on the plasma density

$$n_e < n_c \equiv \frac{m_e \omega^2}{4\pi e^2} = 1.1 \times 10^{21} \text{ cm}^{-3}(\lambda/1\ \mu\text{m})^{-2}, \tag{3.7}$$

where n_c is called the cut-off or, more often, the "critical" density. A plasma having electron density $n_e < n_c$ for a particular wavelength λ, allowing the propagation of transverse EM waves of such wavelength, is said to be *underdense*. In the opposite condition of an *overdense* plasma having $n_e > n_c$, $\varepsilon = 1 - n_e/n_c < 0$, thus n and k assume imaginary values, and the EM wave cannot propagate into the plasma: the field becomes exponentially evanescent as $\sim \exp(-x/\ell_s)$ where

$$\ell_s = c(\omega_p^2 - \omega^2)^{-1/2}. \tag{3.8}$$

Equation (3.7) shows that for typical laser wavelengths $\lambda \sim 1\ \mu\text{m}$ gas targets will be typically underdense and solid targets will be overdense. Essentially, Eq. (3.7) also explains why a metal is a mirror for visible light and why radio waves are reflected by the ionosphere.

When the electron density is inhomogeneous, e.g. $n_e = n_e(x)$, both n and k will be also functions of the spatial coordinate. A limiting case of special interest is that of a step-like density profile, i.e. $n_e = n_0\Theta(x)$ with $\Theta(x) = 0$ for $x < 0$ and $\Theta(x) = 1$

for $x > 0$. In such a case Fresnel formulas (see e.g. Jackson 1998, Sect. 7.3) may be used to evaluate the amplitudes of reflected and transmitted waves from such a step boundary plasma. The latter is a reasonable model for the interaction of an ultrashort laser pulse with a solid target if the plasma expansion before and during the interaction is negligible, so that the density scalelength $L_n = n_e/|\partial_x n_e| \ll \lambda$. In the opposite limit $L_n \gg \lambda$, the propagation of EM waves may be studied using the JWKB approximation (see e.g. Bender and Orszag 1999, Chap. 10).

The wave Eq. (3.4) also allows solutions for *longitudinal, electrostatic* (ES) waves having $\mathbf{V} \times \mathbf{E} = 0$, $\mathbf{V} \cdot \mathbf{E} \neq 0$ and $\mathbf{B} = 0$. For a plane wave this implies $\mathbf{k} \parallel \mathbf{E}$, i.e. the electrons oscillate along the direction of the wavevector. Thus, $\nabla^2 \mathbf{E} = \mathbf{V}(\mathbf{V} \cdot \mathbf{E})$ and we obtain from (3.4) that ES waves exist only for frequencies such that $\varepsilon(\omega) = 0$, i.e. for $\omega = \omega_p$, and that the wavevector k is not determined by the wave equation. These features also imply that the group velocity vanishes, $v_g = \partial_k \omega = \partial_k \omega_p = 0$, and the phase velocity $v_p = \omega_p/k$ may have arbitrary values. These ES waves at the plasma frequency are simply called plasma waves or *plasmons*. They are associated to a charge density perturbation that can be obtained from Poisson's equation

$$\delta \varrho = -e\delta n_e = -e(n_e - Zn_i) = \frac{1}{4\pi} \mathbf{V} \cdot \mathbf{E}, \qquad (3.9)$$

where n_i is the density of background ions of charge Z.

As an aside, we also derive the dispersion relation of plasma waves in a *warm* plasma of electron temperature T_e when the fluid velocity $u_x \ll v_{te} = (T_e/m_e)^{1/2}$, the thermal velocity. To this aim we linearize Eqs. (2.67) and (2.71) for electrons in 1D,

$$\partial_t n_e \simeq -n_0 \partial_x u_x, \qquad \partial_t u_x \simeq -\frac{e}{m_e} E_x - \frac{\gamma_e T_e}{m_e} \frac{\partial_x n_e}{n_0}, \qquad (3.10)$$

where $P_e = \gamma_e n_e T_e$ with $\gamma_e = 3$ is assumed (see Sect. 2.2.2). To close the system we use $\partial_x E_x = 4\pi e(n_e - n_0) = 4\pi e\delta n_e$. Looking for plane wave solutions we obtain

$$-i\omega\delta\tilde{n}_e = -ikn_0\tilde{u}_x, \qquad -i\omega\tilde{u}_x = -\frac{e}{m_e}\tilde{E}_x - 3v_{te}^2\frac{\delta\tilde{n}_e}{n_0}, \qquad ik\tilde{E}_x = 4\pi\delta\tilde{n}_e. \quad (3.11)$$

This linear system has a solution if

$$\omega^2 = \omega_p^2 + 3k^2 v_{te}^2, \qquad (3.12)$$

that is the dispersion relation. The same result can be obtained using the linearized Vlasov equation with a background Maxwellian distribution. However, such kinetic approach shows that the frequency has also an imaginary part, implying that the waves are damped (*Landau damping*). Details may be found in any textbook of plasma physics. Here we remark that (3.12) is not valid for plasmas where $u_x > v_{te}$ which are described by cold fluid equations.

3.2 Relativistic Electromagnetic Waves

Let us now face the problem of the propagation of an EM wave of "relativistic" amplitude, i.e. of such a high intensity that the motion of electrons becomes relativistic. The
relation between the electron fluid velocity $\mathbf{u}_e$ and the fields is thus nonlinear due to
both the effect of the magnetic force and the nonlinear relation between momentum
and velocity, $\mathbf{p}_e = m_e \gamma(|\mathbf{u}_e|)\mathbf{u}_e$. As a consequence, relativistic and magnetic force
effects bring us on the difficult ground of nonlinear optics, where the propagation
and dispersion properties of a wave depend on its amplitude, generation of multiple
frequencies occur, and so on.

There is, however, an example case which allows to establish some features of
"relativistic" propagation in a simple way. For a plane wave with *circular* polarization, there is a solution of Eq. (2.74) for which the $\mathbf{u}_e \times \mathbf{B}$ term vanishes, the
momentum is purely transverse ($\mathbf{p}_e \cdot \mathbf{k} = 0$) and the factor γ_e is constant. (The situation is analogous to that for the single electron in Sect. 2.1.3). In this case the effective
equation of motion may be written as

$$\frac{d}{dt}(m_e\gamma_e\mathbf{u}_e) = m_e\gamma_e\frac{d\mathbf{u}_{e,\perp}}{dt} = -e\mathbf{E}, \qquad \gamma_e = \gamma_e(a_0) = (1 + a_0^2/2)^{1/2}, \quad (3.13)$$

where we stressed that the electrons move in the transverse plane. The equation is
identical to the linear case except for the constant factor γ_e multiplying the electron
mass, so that we can replace $m_e \to m_e\gamma_e$ in Eqs. (3.5) and (3.6) to obtain a nonlinear
version of $\varepsilon(\omega)$, $n(\omega)$ and the dispersion relation as (Akhiezer and Polovin 1956)

$$\varepsilon_{NL}(\omega) = n_{NL}^2(\omega) = 1 - \frac{\omega_p^2}{\gamma_e\omega^2}, \qquad -k^2c^2 + \omega^2 - \frac{\omega_p^2}{\gamma_e} = 0. \quad (3.14)$$

At this point we should warn the reader that the concepts of refractive index and
dispersion relation must be taken with care in the context of nonlinear optics, and that
one should keep in mind that the relations (3.14) has been established for a particular
case. Since Eqs. (3.14) depend on the amplitude of the EM field, the propagation of
any realistic laser pulse, of finite extension both in space and time, will be affected in
general by complicated effects of nonlinear propagation and dispersion which will
modify the spatial and temporal shape of the pulse itself. Even choosing a different
wave polarization may lead to very different results: for instance, if the polarization
is linear it is easily found that the relativistic factor γ_e is not a constant (or at least
a slowly varying function of time) but it is rapidly oscillating and its Fourier series
contains an infinite number of terms; this means that, in general, the propagation of
a linearly polarized EM wave of "relativistic" amplitude will lead to the generation
of high harmonics. Nevertheless, a careful analysis reveals that the propagation of
the first harmonic component, i.e. of the "main" wave, is reasonably described by
(3.14) where $\gamma_e \to \langle\gamma_e\rangle$, the cycle-averaged relativistic factor (see e.g. Sprangle et
al. 1990).

Despite their limited validity Eqs. (3.14) give a sufficient basis to infer the occurrence of two main phenomena induced by relativistic effects, namely *relativistic self-focusing* (Sect. 3.3) and *relativistic transparency* (Sect. 3.4). Indeed, there are several other nonlinear phenomena where the relativistic effect cooperate with that of ponderomotive forces. Examples may be found in review papers (see e.g. Bulanov et al. 2001; Mourou et al. 2006).

3.3 Relativistic Self-Focusing

3.3.1 Simple Model

Equations (3.14) lead to the prediction that an intense laser beam may undergo self-focusing (SF), overcoming the effect of diffraction. Let us consider a laser beam with a typical Gaussian-like profile in the transverse direction, so that the dimensionless amplitude has its peak value a_0 on the axis and decreases with increasing distance r from the axis. The nonlinear intensity dependence of (3.14) implies that the refractive index will be higher on the axis and decrease with r, similarly to what happens in an optical fiber. From a geometrical optics viewpoint, the light rays will be bent and re-collimated producing a converging lens effect. SF will occur when the optical power of such lens overcomes the natural diffraction of the beam.

The optical fiber picture allows to show that there will be a laser *power* threshold for SF and to estimate the scaling and value of such power. Let us consider a light ray coming from the axial region and forming an angle θ_i with respect to the radius, as in Fig. 3.1. The ray propagates from the region near the axis, where the nonlinear refractive index has the value $n_0 = n_{NL}(a_0)$, to the edge region where the field vanishes and the nonlinear refractive index has the value $n_1 = n_{NL}(0) < n_0$. According to Snell's law, the ray will be bent at a refraction angle θ_r such that

$$\frac{\sin \theta_r}{\sin \theta_i} = \frac{n_0}{n_1} > 1, \tag{3.15}$$

Fig. 3.1 Simple "optical fiber" picture of relativistic self-focusing. The beam is confined in the central region of higher refractive index $n_0 > n_1$ when the refraction angle $\theta_r > \theta_i \simeq \lambda/D$, the diffraction angle

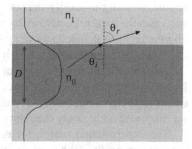

so that *total* reflection occurs when $\sin \theta_r = 1$. We take this condition as the threshold for SF. Assuming relativistic effects to be weak ($a_0 \ll 1$) we can approximate $\gamma = \left(1 + a_0^2/2\right)^{1/2} \simeq 1 + a_0^2/4$, and thus

$$\varepsilon_{\mathrm{NL}} \simeq 1 - \frac{\omega_p^2}{\omega^2}\left(1 - \frac{a_0^2}{4}\right), \qquad \frac{n_0}{n_1} \simeq 1 + \frac{\omega_p^2/8\omega^2}{1 - \omega_p^2/\omega^2}a_0^2. \qquad (3.16)$$

Furthermore we assume angles to be small as in the paraxial approximation ($\theta_i \ll 1$) and take θ_i to be the typical angle of diffraction, i.e. $\cos\theta_i \simeq \lambda/D$ where D is the diameter of the beam. Thus the threshold condition can be written as

$$\frac{n_0}{n_1} := \frac{1}{\sin\theta_i} \simeq \frac{1}{(1 - \lambda^2/D^2)^{1/2}} \simeq 1 + \frac{\lambda^2}{2D^2}. \qquad (3.17)$$

Assuming a low-density, well underdense plasma with $\omega_p^2/\omega^2 \ll 1$ we can further simplify this latter equation obtaining the condition

$$\pi\left(\frac{D}{2}\right)^2 a_0^2 = \pi\lambda^2\frac{\omega^2}{\omega_p^2}, \qquad (3.18)$$

where the l.h.s. is proportional to the product of the beam intensity and the focal area, i.e. to the power of the beam $P = \pi (D/2)^2 I$. By recalling the expression (2.15) giving a_0 in practical units we obtain a threshold power

$$P_c \simeq 4.3 \times 10^{10} \, \mathrm{W} \frac{\omega^2}{\omega_p^2} = 43 \, \mathrm{GW}\frac{n_c}{n_e}, \qquad (3.19)$$

that depends only on the ratio between the plasma density n_e and the cut-off density n_c. This result, based on a very rough modeling, yields the correct scaling and order of magnitude of the commonly accepted value for the critical power that can be obtained from a fairly more rigorous theory for a Gaussian beam (Sect. 3.3.3).

Due to the finite width of the laser beam, there will be a radial component of the ponderomotive force (Sect. 2.1.4) which will expel electrons out of the central region, creating a density depression around the axis. Such density decrease also increases the local refractive index and hence has a focusing effect for the pulse ("self-channeling"). A detailed modeling of self-focusing thus has to take into account both the "pure" relativistic effect, which comes from the effective inertia of electrons, and the self-consistent modification of the density profile, which is determined by the balance of the ponderomotive force with the space-charge electric force produced by charge displacement. However, within suitable approximations the SF threshold may be estimated taking only the relativistic contribution into account, see Sect. 3.3.3.

The experimental search for relativistic SF and self-channeling has been very intense during the 1990s, as multi-terawatt lasers became available. Most important experiments are reviewed in Gibbon (2005) (Sect. 4.8). Such experiments were able to provide evidence, mostly via optical diagnostics, for long and narrow channel formation in an underdense plasma. More recently, using a proton beam as a charged particle probe, it was possible both the resolve in time the propagation of the laser pulse inside the plasma and the formation of a charged channel due to the ponderomotive action (Kar et al. 2007).

3.3.2 Multiple Scale Analysis. Envelope Equation

The development of a rigorous and quite complete theory of relativistic SF requires a substantial analytical effort. Showing a complete account of the calculations is beyond the aim of the present work, but we wish at least to present the theory in a compact way leaving details to the literature. This presentation allows us to briefly introduce some important methods of nonlinear physics and to discuss the basic assumptions of the theory, in order to make its range of validity clear.

Following Sun et al. (1987), the starting point is the system of cold fluid Eqs. (2.67–2.74) plus the wave equations from the EM potentials $\mathbf{A}$ and Φ, such that $\mathbf{E} = -c^{-1}\partial_t \mathbf{A} - \nabla\Phi$ and $\mathbf{B} = \nabla \times \mathbf{A}$, in the Lorentz gauge $\nabla \cdot \mathbf{A} + c^{-1}\partial_t \Phi = 0$. It is convenient to use the dimensionless quantities $\mathbf{a} \equiv e\mathbf{A}/m_e c^2$, $\varphi \equiv e\Phi/m_e c^2$, $n \equiv n_e/n_0$ (with n_0 the background density) and to normalize time to ω^{-1} and space to c/ω. Thus, the wave equations are

$$(\nabla^2 - \partial_t^2)\mathbf{a} = \alpha^2 n\mathbf{v}, \qquad (\nabla^2 - \partial_t^2)\varphi = \alpha^2(n-1), \tag{3.20}$$

where $\alpha^2 = (\omega_p^2/\omega^2) = n_0/n_c$. The gauge condition for the potentials is $\nabla \cdot \mathbf{a} + \partial_t \varphi = 0$, and the cold fluid equations for electrons are

$$d_t(\mathbf{p} - \mathbf{a}) = \nabla\varphi - \mathbf{v} \cdot (\nabla\mathbf{a}), \qquad \partial_t n + \nabla \cdot (n\mathbf{v}) = 0. \tag{3.21}$$

As usual, $\mathbf{p} = m_e \gamma \mathbf{v}$ with $\gamma = (1 + \mathbf{p}^2)^{1/2}$. Notice that the vector $\nabla\mathbf{a}$ is in tensor form, so that the vector components $[\mathbf{v} \cdot (\nabla\mathbf{a})]_i = v_j\partial_i a_j$. Ions are assumed as an immobile background.

The quantity α will be used as the ordering parameter to reduce the above system to a simpler, more compact one. The fundamental assumptions are that the plasma is well underdense, i.e. $\alpha = \omega_p/\omega \ll 1$, and that the laser pulse represented by $\mathbf{a}$ and φ is sufficiently wide and long with respect to the smallest scale at play, i.e the laser wavelength.

To put these assumptions in a rigorous frame, we introduce the method of multiple scales using an example from Bender and Orszag (1999), Chap. 11. Consider the equation of motion for an oscillator with a nonlinear cubic term (Duffing oscillator) and the related initial conditions for the coordinate $q(t)$

$$d_t^2 q = -q - \epsilon q^3, \qquad q(0) = 1, \qquad d_t q(0) = 0, \tag{3.22}$$

where the nonlinear term is assumed to be small, i.e. $\epsilon \ll 1$. In the standard perturbative approach we write $q(t) = q_0(t) + \epsilon q_1(t)$ obtaining, to lowest order, $q_0 = \cos t$, while for the first order term

$$d_t^2 q_1 = -q_1 - \epsilon \cos^3 t = -q_1 - (\epsilon/4)(\cos 3t + 3\cos t). \qquad (3.23)$$

The last term, albeit of small amplitude, is resonant with the natural frequency of the linear oscillator, so that we obtain a solution containing a secular term $\sim t \sin t$ growing with time. This is in contrast with the fact that $q(t)$ is upper limited, since the energy of the system $\mathcal{E} = 1/2 + \epsilon/4$ is a constant (and we may further notice that (3.22) can be obtained from the Hamiltonian $H = p^2/2 + q^2/2 + \epsilon q^4/4$ that is a constant of motion). Thus, the perturbative solution is valid for very small times $t \lesssim \epsilon^{-1}$. To overcome this difficulty we refine the perturbation approach by defining an additional variable $t_1 = \epsilon t$ and look for a solution in the form

$$q(t) = q_0(t, t_1) + \epsilon q_1(t, t_1), \qquad (3.24)$$

where t and t_1 are treated as independent variables, so that $d_t q = \partial_t q + (d_{t_1} t)\partial_{t_1} q$ with $dt_1/dt = \epsilon$. Thus, to order ϵ^2,

$$d_t q = \partial_t q_0 + \epsilon(\partial_{t_1} q_0 + \partial_t q_1), \qquad d_t^2 q = \partial_t^2 q_0 + \epsilon(2\partial_{t_1}\partial_t q_0 + \partial_t^2 q_1), \qquad (3.25)$$

so that the following equations are obtained

$$\partial_t^2 q_0 = -q_0, \qquad \partial_t^2 q_1 = -q_1 - q_0^3 - 2\partial_{t_1}\partial_t q_0. \qquad (3.26)$$

The first equation has a solution of the form $q_0 = Q(t_1)e^{it} + \text{c.c.}$, where the function Q_1 is determined by substituting in the second equation and imposing that the "nasty" resonant term vanishes, yielding $Q = e^{3it_1/8}/2$. Thus, the solution to order ϵ is

$$q(t) = \cos[(1 + 3\epsilon t/8)], \qquad (3.27)$$

where it can be seen that the nonlinear term shifts the oscillation frequency, an effect not found using the simplest perturbation approach. In other words, it is now clear that t_1 describes the nonlinear variations of q on a temporal scale *slower* than the one described by t.

Going back to SF, the idea is to pursue a similar approach based on three different scales: the width and the length of the laser pulse are assumed to be of order $\sim\alpha^{-1}\lambda$ and $\sim\alpha^{-2}\lambda$, respectively, with the laser wavelength λ giving the smallest scale. As suggested by the above example, the multiple scale expansion also allows to keep track of the nonlinear modifications to the dispersion relation of the EM wave. Eventually this approach leads to a set of simplified equations and related solutions characterizing different regimes of SF and providing a derivation of the SF power threshold more accurate than that provided by simple models. However, it must be kept in mind that the ordering assumptions are not valid anymore for either tightly focused or extremely ultrashort pulses, whose width and duration, respectively, are of the order of very few wavelengths (so-called "λ^3" pulses). These pulses are of current research interest and their propagation in a plasma may not be described with the present approach.

Consistently with the above assumptions and also assuming cylindrical symmetry, the vector potential describing the laser pulse (propagating along the x direction) is taken in the form (Sun et al. 1987)

$$\mathbf{a}(\mathbf{r}, t) = \mathbf{a}_0(x_0 - v_f t, x_2, r_{1\perp}) + \alpha \mathbf{a}_1(x_0 - v_f t, x_2, r_{1\perp}), \qquad (3.28)$$

$$\mathbf{a}_0 = \mathrm{Re}\left[\frac{\hat{\mathbf{y}} \pm i\hat{\mathbf{z}}}{\sqrt{2}} e^{ik(x_0 - v_f t)} \tilde{a}(x_2, r_{1\perp})\right], \qquad (3.29)$$

where $x_0 = x$, $r_{1\perp} = \alpha(y, z)$ and $x_2 = \alpha^2 x_0$. The dimensionless phase velocity $v_f = \omega/kc$, which must be determined *a posteriori*, is assumed to be different from unity by terms of order α^2, i.e. $v_f = 1 + \mathcal{O}(\alpha^2)$. The term $\mathbf{a}_0$ describes the transverse components of the laser pulse, the longitudinal components being of higher order. Circular polarization is assumed in order to avoid high-frequency longitudinal oscillations driven by the $\mathbf{v} \times \mathbf{B}$ force. Within these assumptions, a consistent expansion of Eqs. (3.20, 3.21) up to order α^2 leads to the following relations for leading order quantities,

$$\mathbf{p}_0 = \mathbf{a}_0, \qquad \nabla_{1\perp}(\varphi_0 - \gamma_0) = 0, \qquad (3.30)$$

with $\gamma_0 = (1 + \mathbf{a}_0^2)^{1/2} = (1 + |\tilde{a}|^2/2)^{1/2}$. The first relation tells us that the conservation of canonical momentum due to translational invariance in 1D (i.e. for plane waves) approximately holds (to order α) if the transverse profile of the pulse is smooth enough. The second relation tells us that, within the same accuracy, electrostatic and ponderomotive forces in radial direction balance each other.

Proceeding with the calculation, eventually the following equation for the laser envelope is obtained

$$(\nabla_{1\perp}^2 + 2i\partial_{x_2} + \eta)\tilde{a} = \frac{n}{\gamma}\tilde{a}, \qquad (3.31)$$

where we defined

$$\eta = \frac{v_f^2 - 1}{v_f^2 \alpha^2} = \frac{\omega^2 - k^2 c^2}{\omega_p^2}. \qquad (3.32)$$

Notice that from now on we may drop for simplicity the suffixes 1 and 2 for r_1 and x_2, as other variables with different scale do not appear anymore. We must however remember that the original definitions $r_1 = \alpha r$ and $x_2 = \alpha^2 x$ with r and x in units of c/ω, now imply that in the following r is normalized to $\alpha^{-1}(c/\omega) = c/\omega_p$ and x to $\alpha^{-2}(c/\omega) = \alpha^{-1}(c/\omega_p) = c\omega/\omega_p^2$, respectively.

The density n is determined by the following equation

$$n = \Theta(1 + \nabla_\perp^2 \gamma), \tag{3.33}$$

that is obtained by combining the second of Eq. (3.30) with Poisson's equation $\nabla_\perp^2 \varphi = n - 1$. The Θ-function is inserted since n cannot become negative: whenever $1 + \nabla_\perp^2 \gamma < 0$, then $n = 0$ indicating that electron cavitation has occurred.

The quantity η in Eq. (3.31) has the role of an eigenvalue: different solutions associated to different boundary conditions determine the value of η and thus the nonlinear dispersion relation between ω and k. For example, assuming an infinite beam ($\partial_{x_2} \tilde{a} = 0$) with cylindrical symmetry and a parabolic dependence with radius $\tilde{a}(r) \simeq \tilde{a}(0)(1 + \kappa^2 r^2)$ near the axis ($r \to 0$), and unperturbed density ($n = 1$), Eq. (3.31) yields for $r = 0$ the relation $(4\kappa^2 + \eta)\tilde{a}(0) = \tilde{a}(0)/\gamma(0)$ (with $\gamma(0) = (1 + |\tilde{a}(0)|^2/2)^{1/2}$), from which we obtain $\omega^2 - k^2 c^2 = (\omega_p^2/\gamma(0))(1 + 4\kappa^2 \gamma(0))$, recovering the plane wave result for $\kappa = 0$. Sun et al. (1987) performed a numerical analysis of Eqs. (3.31–3.33) characterizing different types of solutions and obtaining values for the SF and cavitation thresholds.

In the literature, the envelope Eq. (3.31) is also sometimes obtained via shorter tracks where some of the assumptions of the rigorous multiple scale approach are implicit. See for example Gibbon (2005), par.4.6.2, where starting from the wave equation for the vector potential (3.20) use is made of the relation $\mathbf{a} = \mathbf{p} = \gamma \mathbf{v}$ that is actually valid only for sufficiently wide beams, i.e for a beam radius larger than $\lambda/\alpha = c/\omega_p$. Such relation is used to write the current density as $\mathbf{J} = -n\alpha^2 \mathbf{a}/\gamma$ with $\gamma = (1 + \mathbf{a}^2)^{1/2}$. Then, use is also made of the *linear* dispersion relation $\omega^2 + k^2 c^2 = \omega_p^2$, that corresponds to taking $\eta = 1$ in (3.32). This procedure leads to the equation

$$(\nabla_\perp^2 + 2i\partial_x)\tilde{a} = \left(\frac{n}{\gamma} - 1\right)\tilde{a}, \tag{3.34}$$

which consistently equals (3.31) with $\eta = 1$. Thus, this derivation actually corresponds to picking up a *particular* class of solutions, which can be used for a fully analytical calculation of the relativistic SF threshold (see next section and Gibbon 2005, par.4.6.2). Fortunately, an extended investigation of (3.31) by numerical methods (Sun et al. 1987) shows that in the limit of uniform density $n = 1$ for which the threshold is evaluated (see next Sect. 3.3.3), also $\eta \to 1$, making the calculation consistent.

3.3.3 Nonlinear Schrödinger Equation. Self-focusing Threshold

Following the discussion at the end of the preceding Sect. 3.3.2, we take $\eta = 1$ and assume a weakly relativistic amplitude $|\tilde{a}| \ll 1$ so that $\gamma \simeq 1 + |\tilde{a}|^2/4$, and a smooth profile such that $\nabla_\perp^2 \gamma \ll 1$, so that the density perturbation is negligible, i.e. $n \simeq 1$. Within these assumptions, Eq. (3.31) becomes

$$i\partial_x \tilde{a} = \frac{1}{2}\nabla_\perp^2 \tilde{a} + \frac{1}{8}|\tilde{a}|^2 \tilde{a}, \tag{3.35}$$

which is a *nonlinear Schrödinger equation* (NLSE), one of the most recurrent and studied equations in nonlinear physics (Sulem and Sulem 1999). The NLSE is equivalent to the Euler-Lagrange equations

$$\frac{\partial \mathcal{L}}{\partial \tilde{a}^*} = \partial_\mu \left(\frac{\partial \mathcal{L}}{\partial(\partial_\mu \tilde{a}^*)} \right), \tag{3.36}$$

where $\partial_\mu = (\partial_x, \nabla_\perp)$ and $\mathcal{L}$ is the functional (also named the Lagrangian *density*)

$$\mathcal{L} = -\frac{i}{2}(\tilde{a}^* \partial_x \tilde{a} - \tilde{a}\partial_x \tilde{a}^*) + \frac{1}{2}\nabla_\perp \tilde{a} \cdot \nabla_\perp \tilde{a}^* - \frac{1}{16}|\tilde{a}|^4. \tag{3.37}$$

The rule is that $\tilde{a}$, $\partial_t \tilde{a}$, $\nabla_\perp \tilde{a}$ and their complex conjugates must be considered as independent variables. It is thus easy to show that

$$\frac{\partial \mathcal{L}}{\partial \tilde{a}^*} = \frac{i\partial_x \tilde{a}}{2} - \frac{\tilde{a}|\tilde{a}|^2}{8}, \qquad \frac{\partial \mathcal{L}}{\partial(\partial_x \tilde{a}^*)} = \frac{i\tilde{a}}{2}, \qquad \frac{\partial \mathcal{L}}{\partial(\nabla_\perp \tilde{a}^*)} = \frac{\nabla_\perp \tilde{a}}{2}, \tag{3.38}$$

so that (3.36) reduces to (3.35). This formulation generalizes the Lagrangian approach to classical mechanics that should be known to most readers, and is strongly used in the quantum theory of fields. Existence of the Lagrangian density $\mathcal{L}$ and of the associated action $S = \int \mathcal{L}d^2rdx$ (where $d^2r = 2\pi r dr$) allows to exploit Noether's theorem (see e.g. Goldstein et al. 2002, Sect. 13.7), stating that for any symmetry operation which leaves S unchanged there is a corresponding *conserved* quantity, e.g. an integral over d^2r that is x-independent (in our case x plays the role of time). Two of such conserved integrals, which will be useful for our case, are

$$H = \int_0^\infty \left(\frac{|\nabla_\perp \tilde{a}|^2}{2} - \frac{|\tilde{a}|^4}{16} \right) d^2r, \qquad P = \int_0^\infty |\tilde{a}|^2 d^2r. \tag{3.39}$$

As the reader may argue H is the Hamiltonian of the system, and its x-independence is equivalent, replacing x by t in the NLSE (3.35), to the conservation of energy for a system whose Lagrangian does not depend explicitly on time. The conservation of P comes instead from "gauge" invariance: the Lagrangian is unaffected if the fields are multiplied by a phase factor $e^{i\varepsilon}$. The quantity P acquires a definite physical meaning for our particular system since it is proportional to the integral of the EM pulse intensity over the beam section, i.e. to the pulse power that is conserved along the beam path.

It is also possible to prove that $dH/dx = 0$ and $dP/dx = 0$ algebraically, of course (see e.g. Gibbon 2005, p.101, for a direct proof that P is a constant). A useful relation that can be also proved in the 2D, cylindrically symmetric case is

$$H = \frac{1}{4} \int_0^\infty r^2 \partial_x^2 |\tilde{a}|^2 d^2 r. \tag{3.40}$$

By exploiting the conservation of H and P the threshold for SF is readily obtained as follows. Defining the beam width $\sigma(x)$ at the position x as an average over the intensity of the beam,

$$\sigma^2(x) = \frac{1}{P} \int_0^\infty r^2 |\tilde{a}|^2 d^2 r, \tag{3.41}$$

we obtain for the width variation with x

$$d_x^2 \sigma^2(x) = \frac{1}{P} \int_0^\infty r^2 \partial_x^2 |\tilde{a}|^2 d^2 r = \frac{4H}{P}, \tag{3.42}$$

where (3.39) have been used. The threshold corresponds to a self-guiding condition for which the width does not vary with x, i.e. $\sigma(x) = \sigma(0)$, that yields $H = 0$. We thus evaluate H at the waist position $x = 0$ there assuming a Gaussian intensity distribution $\tilde{a}(r, 0) = a_0 \exp(-r^2/\sigma_0^2)$, obtaining

$$H = \frac{\pi a_0^2}{2} \left(1 - \frac{a_0^2 \sigma_0^2}{16} \right). \tag{3.43}$$

Posing $H := 0$ yields $a_0^2 \sigma_0^2 = 16$. To recover physical units for the pulse power $P = \pi a_0^2 \sigma_0^2$ we recall that the intensity $I = m_e c^3 n_c a_0^2/2$, with $n_c = m_e \omega^2/(4\pi e^2)$ the cut-off density, and that σ_0 is normalized to c/ω_p. Thus, $H = 0$ and self-guiding occurs when the real pulse power attains the "critical" value

$$P_c = 16\pi \frac{m_e^2 c^3 \omega^2}{4\pi e^2} \left(\frac{c}{\omega_p} \right)^2 = 2 \frac{m_e c^3}{r_c} \left(\frac{\omega}{\omega_p} \right)^2 = 17.5 \left(\frac{\omega}{\omega_p} \right)^2 \text{ GW}. \tag{3.44}$$

For $P > P_c$ and $P < P_c$ the beam is expected to self-focus and to diffract, respectively. Equation (3.44) can be considered a refinement of the simple estimate (3.19). However it is important to keep in mind the several assumptions made to obtain this result: a Gaussian beam satisfying the ordering defined by Eq. (3.28), a particular choice of boundary conditions yielding Eq. (3.34), a weak relativistic nonlinearity and an uniform density.

3.4 Relativistic Transparency

According to Eq. (3.14) a plane, monochromatic, circularly polarized wave in a homogeneous plasma can propagate in regions for which $\omega > \omega_p/\gamma^{1/2}$, i.e. up to electron densities $n_e = n_c\gamma$. This increase of the effective cut-off density $n'_c = n_c\gamma > n_c$ with respect to the linear case is known as relativistic *self-induced transparency* (SIT) or, briefly, relativistic transparency.

It should be kept in mind that, due the nonlinear character of Eq. (3.14), in realistic conditions the dynamics of SIT is complicated. For any real pulse having a finite profile and attempting to penetrate inside a plasma with density $n_e > n_c$, the "portions" of the pulse for which the local amplitude is such that $n_c\gamma > n_e$ may propagate while other "portions" are unable to, leading to a shaping of the pulse.

Moreover, the penetration (or reflection) of a wave inside the overdense plasma depends on the initial and boundary conditions. In the simplest scenario, a laser pulse impinges on a step-like plasma with an homogeneous profile extending, say, for $x > 0$, i.e. $n_e = n_0\Theta(x)$. The ponderomotive force will push electrons at the surface of the plasma, creating a charge separation layer and piling up electrons in some region where the density will increase with respect to the initial value n_0. Such situation is sketched in Fig. 3.2a. Thus, the propagation of the laser pulse must be evaluated self-consistently with the density profile modification.

In the following we describe an approach to SIT based on steady, plane wave solutions in the limiting cases of a semi-infinite plasma (Sect. 3.4.1) and of an ultrathin plasma slab (Sect. 3.4.2). Of course, experimental scenarios are always much more complicated than such two reference cases. The targets often may not be considered to be either thick or thin, and the effects of heating and rarefaction due to plasma expansion both in the longitudinal and transverse direction also play an important

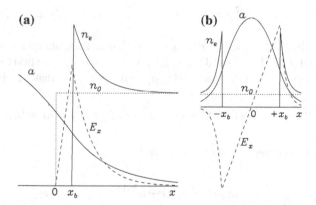

Fig. 3.2 **a** Sketch of the field and density profiles for the nonlinear penetration of a plane wave inside an overdense plasma extending for $x > 0$ (a: EM vector potential, *thin solid line*; n_e: electron density, *thick solid line*; n_0: background density, *dotted line*; E_x: electrostatic field, *dashed line*). **b** The symmetrized solution describing an electromagnetic "caviton"

role. These issues may partly explain the difficult of "clean" observation of SIT in experiments. Early experimental investigations are described, e.g., by Fuchs et al. (1998). Very recently, SIT has been investigated with ultrafast (sub-picosecond) temporal resolution in expanding foils, providing indications of the transition to transparency that acts as a nonlinear optical shutter for the incident pulse (Palaniyappan et al., 2012).

3.4.1 Semi-Infinite Plasma

Let us consider a step-like plasma with initial density profile $n_e = n_0 \Theta(x)$ and a plane wave at normal incidence and with circular polarization (CP), so that we avoid high-frequency longitudinal oscillations. As sketched in Fig. 3.2, the ponderomotive push creates a pile-up of electrons and a charge depletion layer in the $0 < x < b$ region, whose extension is determined by a balance between electrostatic and ponderomotive forces. This regime of nonlinear penetration has been discussed by several authors (Marburger and Tooper 1975; Cattani et al. 2000; Goloviznin and Schep 2000) using the relativistic cold fluid Eqs. (2.67–2.74) plus the wave equations for the potentials $\mathbf{a}$ and φ as in Sect. 3.3.2 in a planar geometry, so that all quantities depend on (x, t). Following Cattani et al. (2000), we look for a *steady* solution such that the electrons oscillate in the transverse plane (with $\mathbf{p}_\perp = \mathbf{a}$) and there is no motion along x, i.e. $p_x = 0$. This is consistent with assuming a standing CP wave $\mathbf{a}(x, t) = a(x)(\hat{\mathbf{y}} \cos \omega t + \hat{\mathbf{z}} \sin \omega t)$ and an equilibrium between the longitudinal field and the ponderomotive force, $eE_x(x) = -e\partial_x \Phi(x) = f_p(x)$. Thus, inside the plasma we are left with the equilibrium condition plus Poisson's equation for $\phi(x)$ and Helmholtz's equation for $a(x)$,

$$d_x(\varphi - \gamma) = 0, \qquad d_x^2 \varphi = n_e - n_0, \qquad d_x^2 a + (1 - n_e/\gamma)a = 0, \qquad (3.45)$$

where we switched to the total derivatives since all quantities are now time independent. Notice that $\gamma = (1 + a^2)$ and n_e, n_0 are normalized to n_c. The first two relations allow to eliminate n_e from the third, using $n_e = n_0 + d_x^2 \gamma$, so that we obtain

$$d_x^2 a - \frac{a}{1 + a^2}(d_x a)^2 + (1 + a^2 - n_0(1 + a^2)^{1/2})a = 0. \qquad (3.46)$$

This equation can be integrated once to obtain

$$\frac{1}{2(1 + a^2)}(d_x a)^2 - n_0(1 + a^2)^{1/2} + \frac{a^2}{2} = -n_0, \qquad (3.47)$$

where we used the assumption that $a(+\infty) = d_x a(+\infty) = 0$ to fix the integration constant. Actually, we can find a Lagrangian $\mathcal{L} = \mathcal{L}(a, d_x a)$ also for Eq. (3.46), following the same approach of Sects. 3.3.3 and Eq. (3.47) turns out to be the related Hamiltonian.

We assume that the electrons occupy the $x > x_b$ region leaving a depletion layer[1] in $0 < x < x_b$, where the parameter x_b will be determined self-consistently as follows. Let $a_<(x)$ and $a_>(x)$ be the solutions for $a(x)$ in the vacuum ($x < x_b$) and plasma ($x > x_b$) regions, respectively. In the vacuum region $x < x_b$, since total reflection occurs we have a standing wave $a_<(x) = 2(a_0/\sqrt{2}) \sin(x+\delta)$ where δ must be determined by the boundary conditions. Defining $a_b \equiv a(x_b)$ and $a_b' = a'(x_b)$, we have $a_b^2 + a_b'^2 = 2a_0^2$. Notice that a_b and a_b' are well defined since both a and $a' \equiv d_x a$ are continuous at $x = b$. Now, the longitudinal field $E_x = -\partial_x \varphi = n_0 x$ for $0 < x < b$, $E_x = -\partial_x \gamma$ for $x > x_b$, and is also continuous at $x = x_b$. Thus, $n_0 x_b := -\partial_x \gamma|_{x=b} = -a_b a_b'/\gamma_b$, yielding $x_b = (a_b/n_0\gamma_b)(2a_0^2 - a_b^2)^{1/2}$.

In the $x > x_b$ region, a solution to (3.47) can be found by switching to a variable u such that $a = 2u/(u^2 - 1)$. After some algebra we find $(d_x u)^2 = (n_0 - 1)u^2 - n_0$ that has a solution $u = n_0^{1/2}\kappa \cosh(\kappa(x - x_0))$ with $\kappa = (n_0 - 1)^{1/2}$, i.e. in physical units $\kappa = \ell_s^{-1}$, see Eq. (3.8). Thus

$$a_>(x) = \frac{2n_0^{1/2}\kappa \cosh(\kappa(x - x_0))}{n_0 \cosh^2(\kappa(x - x_0)) - n_0 + 1}. \tag{3.48}$$

The free parameter x_0 must be chosen to match the vacuum solution for $x < x_b$. Posing $a_>(x_b) := a_b$ and $a_>'(x_b) := a_b'$ in (3.48) and using the above expression for x_b, two implicit equations for a_b and x_0 as functions of a_0 and n_0 are obtained, determining the full solution. A necessary condition for the evanescent solution, however, is that $d_x \gamma < 0$, ensuring that electrons are pushed inside the target. This is equivalent to require that $n_e \geq 0$ or $x \geq x_b$. Using $n_e = n_0 + d_x^2 \gamma$, the wave Eqs. (3.45) and (3.47), one obtains $n_e = \gamma(3\gamma n_0 - 2(n_0 + a^2))$. The limiting condition $n_e(x_b) := 0$ at the boundary, where a has the maximum value inside the plasma region, is taken as the SIT threshold. The latter is obtained as an implicit relation between n_0 and a_0 from $3\gamma_b n_0 - 2(n_0 + a_b^2) = 0$ and expressing a_b as a function of a_0. In the limit $n_0 \gg 1$, one finds $a_b \simeq 3n_0/2$ and thus the condition for SIT

$$a_0 > (3^{3/2}/2^3)n_0^2 \simeq 0.65n_0^2 \quad (n_0 \gg 1), \tag{3.49}$$

which is a much more restrictive condition than $\gamma \simeq a_0 > \sqrt{2}n_0$ as would be suggested by (3.14). When (3.49) holds, the incident pulse penetrates into the plasma as a nonlinear wave.[2] In realistic and/or more general cases, the penetration dynamics may become complicated because of several factors not included in the above model,

---

[1] Actually a depletion region appears only if $n_0 \geq 1.5$ (in units of n_c) (Cattani et al. 2000). For the sake of brevity we restrict ourselves to the $n_0 > 1.5$ case; a more complete discussion can be found in the references.

[2] In Goloviznin and Schep (2000) it was also shown that in some range of parameters evanescent and propagating solutions may coexist, allowing in principle for instability and hysteresis effects, although only the evanescent solution is observed in simulations.

such as e.g. pulse profile or electron oscillations and heating occurring in particular for linear polarization.

3.4.2 Ultrathin Plasma

A simple and useful example of a steady solution can be found for a very thin plasma foil of thickness ℓ whose density has a Dirac delta-like profile (Vshivkov et al. 1998), i.e. $n_e(x) = n_0\ell\delta(x)$. Still we assume the incident plane wave to be monochromatic and circularly polarized so that electron move only in the $x = 0$ plane. In such foil the current is localized at $x = 0$ and the magnetic field is thus discontinuous, as in a perfect mirror, with a jump given by $B(0^+,t) - B(0^-,t) = (4\pi c)\iota(t)$ where $\iota(t) = J(t)\ell$ is the surface current. The electric field E and thus the vector potential A are continuous, and using $p = eA/c$ we can write

$$J = -en_0 v = -en_0 p/(m_e\gamma) = -e^2 n_0 A(0,t)/(m_e\gamma c). \qquad (3.50)$$

Writing the dimensionless vector potential as $a(x,t) = \mathrm{Re}[a(x)e^{-i\omega t}]$, the boundary conditions are

$$a(0^+) - a(0^-) = 0, \qquad \partial_x a(0^+) - \partial_x a(0^-) = -\frac{\omega^2}{c^2}\frac{n_0}{n_c}\ell\frac{a(0)}{\gamma(0)}, \qquad (3.51)$$

where $\gamma(0) = (1 + |a(0)|^2/2)^{1/2}$. Considering the wave incident from the $x < 0$ region the general solution can be written as

$$a(x) = \begin{cases} a_0 e^{ikx} + a_r e^{-ikx} & (x < 0), \\ a_t e^{ikx} & (x > 0), \end{cases} \qquad (3.52)$$

so that the boundary conditions (3.51) yield $a_0 + a_r - a_t = 0$ and $a_0 - a_r - a_t = -2i\zeta a_t(1 + |a_t|^2/2)^{-1/2}$, where

$$\zeta = \frac{\omega_p^2\ell}{kc^2} = \pi\frac{n_0\ell}{n_c\lambda}. \qquad (3.53)$$

By eliminating a_r and noticing that we may take a_0 as a real number, we obtain $a_0^2 = |a_t|^2\left(1 + \zeta^2/(1 + |a_t|^2/2)\right)$ which can be solved for $|a_t|^2$. The real root is

$$|a_t|^2 = \sqrt{\left(1 + \zeta^2 - \frac{a_0^2}{2}\right)^2 + 2a_0^2} - \left(1 + \zeta^2 - \frac{a_0^2}{2}\right) \equiv Ta_0^2 \equiv (1 - R)a_0^2,$$

$$(3.54)$$

where $T = |a_t|^2$ and $R = |a_r|^2$ are the reflection and transmission coefficients according to their usual definition. The relation $T + R = 1$ can be verified by noticing that $|a_r|^2 = |a_t|^2 + a_0^2 - 2a_0 \text{Re}(a_t)$ and $\text{Re}(a_t) = |a_t|^2/a_0$.

For $\zeta > 1$, a possibly useful approximation of the exact formula for R is given by

$$R = \begin{cases} \zeta^2/(1+\zeta^2) & (a_0 < \sqrt{\zeta^2 - \zeta^{-2}}), \\ (\zeta^2 - 1)/a_0^2 & (a_0 > \sqrt{\zeta^2 - \zeta^{-2}}), \end{cases} \tag{3.55}$$

showing that R decreases abruptly with increasing a_0 beyond a transparency threshold defined as

$$a_0 = \sqrt{\zeta^2 - \zeta^{-2}} \simeq \zeta, \tag{3.56}$$

where the last equality holds for $\zeta \gg 1$. The SIT threshold for a thin foil thus depends on its surface density $n_0 \ell$. The approximation of a Dirac delta profile is rigorously justified if $\ell \ll \ell_s = c/\omega_p$, but numerical simulations show that (3.56) fairly holds even for larger $\ell \lesssim \lambda$ (see e.g. Macchi et al. 2009).

Present-day manufacturing technology allows to produce ultrathin solid foils down to $\ell \simeq 10^{-2}\lambda$, so that in principle (3.56) is experimentally accessible, although in realistic conditions the pulse transmission will be also affected by the target bending and rarefaction since the pulse width will be finite. The regime of interaction with such ultrathin foils is of present interest for laser-driven ion acceleration, and related experiments have provided indications of the onset of SIT (see Chap. 5).

3.5 Electromagnetic Cavitons and Post-solitons

Let us now consider again the solution of Sect. 3.4.1, which is determined by matching the vacuum side solution, $a_<(x) = 2(a_0/\sqrt{2})\sin(x + \delta)$, with $a_>(x)$ given by (3.48) and describing an evanescent wave inside the plasma. The boundary conditions determine the parameters δ and x_0 as a function of a_0 and n_0 (provided that the SIT threshold is not exceeded). Now, assume as an additional condition $\delta = \pi/2$ that determines a particular solution existing only for a well defined choice of a_0 and n_0, which are *not* independent anymore. For this solution, $a(x)$ has a maximum in $x = 0$. Let us take this solution for $x > 0$, and extend it for $x > 0$ with the symmetry relations $a(-x) = a(x)$, $n_e(-x) = n_e(x)$, and $E_x(-x) = -E_x(x)$, as in Fig. 3.2b. The condition $\delta = \pi/2$ ensures that both $a(x)$ and $\partial_x a(x)$ are continuous at $x = 0$. This solution describes a standing EM wave *trapped* inside a cavity in the electron density, extending for $x < |x_b|$, and whose walls are sustained by the internal radiation pressure against the electrostatic field. Such structure may be called an electromagnetic *caviton* or *soliton*[3] being localized in space.

---

[3] The second definition is by far the most used, although the present author is not fully satisfied with that as he believes the term "soliton" to identify a more specific nonlinear structure.

EM structures of such type have been often observed in numerical simulations, typically appearing during turbulent phases of nonlinear propagation or in plasmas with n_e not far from n_c. Qualitatively, the route to the formation of such structures may be described considering the local action of the ponderomotive force, which creates small-scale density depressions where the lower frequency components of the EM pulse may be trapped. Furthermore, the frequency may decrease due to nonlinear effects (see e.g. the example in Sect. 3.3.2), favoring the self-trapping of the field. The simulations have also highlighted examples of the complex EM structure in a "real" 3D geometry (Esirkepov et al. 2002). Reviews of theoretical and numerical studies may be found in the literature (see e.g. Bulanov et al. 2001; Mourou et al. 2006).

Even in the highly idealized 1D above described configuration, EM cavitons are stable only as long as the ions are considered as immobile. On a typical time scale $\sim\omega_{pi}^{-1}$, with $\omega_{pi} \sim (m_p/m_e)\omega_p$ the ion plasma frequency, the electrostatic field accelerates the ions and drive an expansion of the cavity. The late time remnants of such evolution, commonly known as *post-solitons*, have been observed experimentally using charged particle probes (see e.g. Romagnani et al. 2010 and references therein).

3.6 Electrostatic Waves

3.6.1 Wake Waves

As mentioned in (3.1), electrostatic (ES) waves in a cold plasma have a simple dispersion relation $\omega = \omega_p$ that does not determine the wavevector k. Hence, the wavelength and phase velocity of the plasma wave are determined by the way the wave is excited. A possibility is to excite an electron oscillation by a force traveling in the plasma at velocity v_f; as the oscillation is produced at the force front, the phase velocity of the wave will be equal by construction to the velocity of the force perturbation. Causality dictates that the plasma oscillations will be excited behind the traveling front, and thus will appear as a *wake* of plasma oscillations having phase velocity $v_p = \omega_p/k = v_f$. A simple expression for the wake in the limit of small oscillations and non-relativistic electron velocity may be written in the case the force is a short pulse delivering a net momentum to electrons in a time shorter than the oscillation period, so that in a plane geometry we may write $f \simeq m_e u_0 \delta(t - x/v_f)$. Such force produces a wake with the electron velocity given by

$$u_x = u_0 \cos\left(\omega_p \tau\right) \Theta(\omega_p \tau), \qquad \tau = t - x/v_f. \tag{3.57}$$

From the linearized fluid equations $\partial_t n_e \simeq -n_0 \partial_x u_x$ and $\partial_t u_x \simeq -eE_x/m_e$ we also obtain the expressions for $\delta n_e = n_e - n_0$ and E_x:

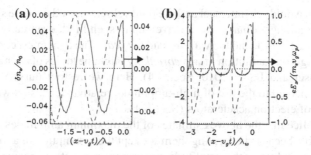

Fig. 3.3 Sheet model simulation (Sect. 2.3.1) where a force pulse travels from left to right at velocity v_g leaving behind a plasma wake with wavelength $\lambda_w = 2\pi v_g/\omega_p$. Frame **a** shows a low amplitude wake with sinusoidal profiles of $\delta n_e/n_0$ (*thick line*) and E_x (*dashed line*). Frame **b** shows a high amplitude wake with a spiky profile of δn_e and a sawtooth profile of E_x. Courtesy of P. Dell'Osso

$$E_x \simeq \frac{m_e \omega_p u_0}{e} \sin\left(\omega_p \tau\right) \Theta(\tau), \qquad \delta n_e \simeq n_0 \frac{u_0}{v_f} \cos\left(\omega_p \tau\right) \Theta(\tau). \qquad (3.58)$$

Wake generation can be studied numerically using the Dawson sheet model (Sect. 2.3.1) as shown in Fig. 3.3a. The wake may also be excited by giving to the first sheet an initial velocity, and the drag effect by the wake field may be studied as a model for the collisionless stopping of a fast particle in the plasma.

As the amplitude of the perturbation grows, the wave undergoes steepening and the profiles of δn_e and E_x appear as shown in Fig. 3.3b. The deformation of the wave may be intuitively understood as the effect of the nonlinear term $\sim u_x \partial_x u_x$ in the 1D fluid Eq. (2.71) causing the crests of the velocity profile to "move faster" than other portions of the wave. Hence, the profiles of u_x and E_x progressively assume a sawtooth shape, that in turn leads (via the continuity equation) to a spiky profile for δn_e, with peak values much greater than n_0.

3.6.2 Relativistic Electrostatic Waves. Wave Breaking

Since the total electron density is a positive quantity, in a plasma wave the density perturbation is limited by the condition $n_e = n_0 + \delta n_e \geq 0$, i.e. $|\delta n_e| \leq n_0$. If we insert this bound in the Poisson's equation $\partial_x E_x = -4\pi e \delta n_e$ and in the *linearized* equation of motion $m_e \partial_x u_x = -e E_x$, and use $k = \omega_p/v_p$ considering the phase velocity v_p as a parameter, we find

$$|E_x| \leq \frac{m_e \omega_p v_p}{e}, \qquad |u_x| \leq v_p, \qquad 2\pi \frac{u_x}{\omega_p} \leq \lambda_p. \qquad (3.59)$$

The third inequality has a clear meaning: in a longitudinal wave, the oscillation amplitude of the particles cannot exceed one wavelength, otherwise the trajectories of the particles would self-intersect and the fluid velocity would become singular (see below). When a longitudinal wave is driven up to such limit, eventually the regular periodic structure is lost, and the wave is said to *break*. The dynamics, consequences, and even the rigorous definition of *wavebreaking* might be topics of a long discussion (see e.g. Mulser and Bauer 2010, Sect. 4.4). Here we restrict ourselves to issues which are connected to the use of high-amplitude, relativistic plasma waves for the development of electron accelerators (Sect. 4.1).

First we notice that the above estimates of the maximum amplitudes of the fields are questionable because in deriving them we kept the assumption of a linear, monochromatic wave, while as the amplitude grows nonlinear effects become important as discussed above (Sect. 3.6.1). Moreover, as will be further clear when discussing electron acceleration in plasma waves (Sect. 4.1), an electron moving with velocity $\simeq \upsilon_p$ may be accelerated to high energy from the wave, since in its rest frame the field is almost constant. This suggests that wavebreaking eventually leads to the wave quickly losing its energy into electrons.

Because of the important role of wavebreaking in electron acceleration and the obvious interest in driving electrons up to high energy, we discuss, following (Akhiezer and Polovin, 1956), the structure and maximum amplitude of relativistic ES plasma wave. We start from Eq. (2.74) in planar geometry plus Maxwell's equations for the ES field,

$$(\partial_t + u_x \partial_x) p_x = -eE_x, \quad \partial_x E_x = 4\pi e(n_0 - n_e), \quad \partial_t E_x = 4\pi e n_e u_x, \quad (3.60)$$

where $p_x = m_e \gamma(u_x) u_x$. We look for a propagating wave solution, i.e. $E_x = E_x(x - \upsilon_p t)$ and so on. It is thus convenient to use $\xi = x - \upsilon_p t$ as an independent variable. For a coordinate transformation $(x, t) \rightarrow (\xi, \tau)$ with the trivial identity $\tau = t$, the ∂_x and ∂_t operators transform as

$$\partial_x = \partial_x \xi \partial_\xi + \partial_x \tau \partial_\tau = \partial_\xi, \quad \partial_t = \partial_t \xi \partial_\xi + \partial_t \tau \partial_\tau = -\upsilon_p \partial_\xi + \partial_\tau, \quad (3.61)$$

where the derivative ∂_τ with respect to τ is effectively a null operator on functions which depend on ξ only. Thus

$$(u_x - \upsilon_p) \partial_\xi p_x = -eE_x, \quad \partial_\xi E_x = 4\pi e(n_0 - n_e), \quad -\upsilon_p \partial_\xi E_x = 4\pi e n_e u_x. \quad (3.62)$$

Eliminating $\partial_\xi E_x$ we can write for n_e

$$n_e = \frac{n_0}{1 - u_x/\upsilon_p} \quad (3.63)$$

which tells us that the density becomes singular when u_x equals the phase velocity, as we inferred above. Differentiating the equation of motion with respect to ξ, we also eliminate $\partial_\xi E_x$ to obtain

$$\partial_\xi[(u_x - v_p)\partial_\xi p_x] = -4\pi e^2 n_0 \left(1 - \frac{v_p}{v_p - u_x}\right) = m_e \omega_p^2 \frac{u_x}{v_p - u_x}. \tag{3.64}$$

Since ξ is the only variable we may replace ∂_ξ with $d_\xi = d/d\xi$. It is also convenient to switch to dimensionless quantities $\xi \to (\omega_p/c)\xi = \zeta$, $u_x \to u_x/c = u$, $v_p \to v_p/c = \beta_p$, and $E = eE_x/(m_e\omega_p c)$. Then, since $p_x = m_e c\gamma(u)u$,

$$d_\zeta[(u - \beta_p)d_\zeta(\gamma u)] = \frac{u}{\beta_p - u}. \tag{3.65}$$

Now, from $(u_x - v_p)\partial_\xi p_x = -eE_x$ and by using some straightforward algebra one can prove

$$E(\zeta) = (u - \beta_p)d_\zeta(\gamma u) = d_\zeta[\gamma(1 - \beta_p u)], \tag{3.66}$$

[hint: first show that $d_\zeta(\gamma u) = \gamma^3 d_\zeta u$ and that $d_\zeta(\gamma) = u d_\zeta(\gamma u)$; remember that $\gamma = (1 - u^2)^{-1/2}$]. Then (3.65) becomes

$$d_\zeta^2[\gamma(1 - \beta_p u)] = \frac{u}{\beta_p - u}. \tag{3.67}$$

The second derivative on the l.h.s. makes us wonder if we may recast this equation in a form similar to Newton's equation in a potential, and this is indeed the case. In fact, defining a "coordinate" $X = X(\zeta) = \gamma(1 - \beta_p u)$, we have $dX = (1 - \beta_p u)d\gamma - \gamma\beta_p du$. By differentiating $u^2 = 1 - \gamma^{-2}$ we obtain $udu = \gamma^{-3}d\gamma$ and then $dX/d\gamma = (u - \beta_p)/u$. Thus the equation for $X(\zeta)$ is

$$\frac{d^2 X}{d\zeta^2} = -\frac{d\gamma}{dX}. \tag{3.68}$$

By integrating between ζ_0 and ζ we obtain for the "velocity" $E = dX/d\zeta$

$$\frac{1}{2}E^2(\zeta) = \gamma_m - \gamma[u(\zeta)], \qquad \gamma_m = \max(\gamma), \tag{3.69}$$

where we used the fact that $E = 0$ when the "potential energy" γ has its maximum value. Since the limiting condition for the wave to exist is $u_x = v_p$, i.e. $u = u_p$, then $\gamma_m = (1 - \beta_p^2)^{-1/2}$. The maximum value of the electric field corresponds to $u = 0$, since the force and the momentum are out of phase for a periodic solution. Thus, posing $\gamma = 1$ and recovering physical units we eventually obtain (Akhiezer and Polovin 1956)

$$\max(E_x) = \frac{m_e \omega_p c}{e}\sqrt{2(\gamma_p - 1)} \equiv E_{wb}, \tag{3.70}$$

which is referred to as the "relativistic wavebreaking" limit. When $\beta_p = v_p/c \ll 1$, one finds again $\max(E_x) = m_e\omega_p v_p/e$.

As the wave amplitude approaches the breaking threshold, the electric field $E_x \propto (\gamma_p - \gamma(u_x))^{1/2}$ acquires a sawtooth shape, with $E_x = 0$ both at the density spikes, i.e where $u_x = v_p$ and n_e diverges, and at the midpoints between the spikes, where $u_x = -v_p$ and n_e has its minima (see Fig. 3.3b). Being $n_e = n_0/(1 - u_x/v_p)$ we thus find $\min(n_e) = n_0/2$.

It is then also possible to estimate the wavelength λ_p of the nonlinear wave (i.e. the distance between adjacent spikes) close to the breaking threshold. As the wave becomes singular, each spikes contains a number of electrons per unit surface $N_e \simeq (n_0/2)\lambda_p$. At the singularity E_x is discontinuous, acquiring the value $\pm E_{\rm wb}$ on each side of the spike; at the same time, integrating Poisson's equation across the spike one finds $E_{\rm wb} = 2\pi e N_e \simeq \pi e n_0 \lambda_p$, from which we obtain

$$\lambda_p \simeq 4(c/\omega_p)\sqrt{2(\gamma_p - 1)}. \tag{3.71}$$

These estimates will be useful in the discussion of strongly nonlinear plasma waves as "flying mirrors" (Sect. 6.2).

References

Akhiezer, A.I., Polovin, R.V.: Sov. Phys. JETP **3**, 696 (1956)
Bender, C.M., Orszag, S.A.: Advanced Mathematical Methods for Scientists and Engineers I: Asymptotic Methods and Perturbation Theory. Springer, Berlin (1999)
Bulanov, S.V., et al.: In: Shafranov, V.D. (ed.) Reviews of Plasma Physics, vol. 22, p. 227. Kluwer Academic/Plenum Publishers, New York (2001)
Cattani, F., Kim, A., Anderson, D., Lisak, M.: Phys. Rev. E **62**, 1234 (2000)
Esirkepov, T., Nishihara, K., Bulanov, S.V., Pegoraro, F.: Phys. Rev. Lett. **89**, 275002 (2002)
Fuchs, J., et al.: Phys. Rev. Lett. **80**, 2326 (1998)
Gibbon, P.: Short Pulse Laser Interaction with Matter. Imperial College Press, London (2005)
Goldstein, H., Poole, C.P., Safko, J.L.: Classical Mechanics. Addison-Wesley, New York (2002)
Goloviznin, V.V., Schep, T.J.: Phys. Plasmas **7**, 1564 (2000)
Jackson, J.D.: Classical Electrodynamics, 3rd edn. Wiley, New York (1998)
Kar, S., et al.: New J. Phys. **9**, 402 (2007)
Marburger, J.H., Tooper, R.F.: Phys. Rev. Lett. **35**, 1001 (1975)
Macchi, A., Veghini, S., Pegoraro, F.: Phys. Rev. Lett. **103**, 085003 (2009)
Mourou, G.A., Tajima, T., Bulanov, S.V.: Rev. Mod. Phys. **78**, 309 (2006)
Mulser, P., Bauer, D.: High Power Laser-Matter Interaction. Springer, Berlin (2010)
Palaniyappan, S., et al.: Nat. Phys. **8**, 763 (2012)
Romagnani, L., et al.: Phys. Rev. Lett. **105**, 175002 (2010)
Sprangle, P., Esarey, E., Ting, A.: Phys. Rev. Lett. **64**, 2011 (1990)
Sulem, C., Sulem, P.: The Nonlinear Schrödinger Equation: Self-Focusing and Wave Collapse. Springer, Applied Mathematical Sciences (1999)
Sun, G.Z., Ott, E., Lee, Y.C., Guzdar, P.: Phys. Fluids **30**, 526 (1987)
Vshivkov, V.A., Naumova, N.M., Pegoraro, F., Bulanov, S.V.: Phys. Plasmas **5**, 2727 (1998)

Chapter 4
Electron Acceleration

Abstract In this chapter we discuss the generation of high-energy electrons in laser-plasma interactions, in two very different regimes. First, we consider electron acceleration in wake waves generated in underdense plasmas, which is the concept behind the development of laser-plasma electron accelerators for high energy physics. Second, we consider the case of an overdense plasma, where electrons are accelerated at the interface where the laser impinges. Such problem is strongly connected to the general issue of collisionless absorption in an overdense plasma, possibly the most complex and less understood topic of laser-plasma interactions.

4.1 Underdense Plasmas: Laser Wakefield Accelerators

4.1.1 Wakefield Generation

How does a surfer gain velocity from a sea wave? It is necessary for the surfer to have some initial velocity along the direction of propagation of the wave. If possible, the initial velocity should be equal to the phase velocity of the wave, because in such a case in its rest frame the surfer would see a constant field which can provide a net acceleration, in contrast to an oscillating field for which the acceleration averages to zero over a cycle. Of course, the process of acceleration has a negative feedback because a change in the surfer's velocity leads to dephasing from the wave.

The above example brings us to two necessary conditions for an electromagnetic wave to accelerate efficiently a charged particle: there must be an electric field component along the propagation direction, and the phase velocity must be such that an optimal phasing between the wave and the particle can occur. Electron plasma waves thus appear as a serious candidate for particle acceleration. In fact, those are longitudinal waves, and the phase velocity v_p is not determined by the plasma frequency, leaving hope to "construct" a wave with the optimal value of v_p. The other attractive point with plasma waves is that electric fields much higher than in conventional

A. Macchi, *A Superintense Laser-Plasma Interaction Theory Primer*,
SpringerBriefs in Physics, DOI: 10.1007/978-94-007-6125-4_4,
© The Author(s) 2013

accelerators may be attained, for the trivial reason that there is no risk of electrical breakdown in an already ionized medium!

Since the final goal is to accelerate relativistic particles, the issue is to generate a plasma wave with phase velocity close to (but not exceeding) that of light, $v_p \lesssim c$, so that relativistic particles (electrons, for definiteness) may remain in phase with the wave. Notice that for highly relativistic energies the acceleration process may be more efficient because a large change in energy corresponds to a small change in velocity, thus the particle may get out of phase only after a long time.

The original work on the "laser electron accelerator" by Tajima and Dawson (1979) proposed to use a wakefield (Sect. 3.6.1) generated by an intense laser pulse[1] propagating at the group velocity $v_g = c(1 - \omega_p^2/\omega^2)^{1/2} \lesssim c$ in an underdense plasma, and exerting on electrons a ponderomotive force (PF) in the longitudinal direction (Sect. 2.1.4)

$$f_p = f_p(x - v_g t) = -m_e c^2 \partial_x \gamma_a, \qquad \gamma_a = \left(1 + \left\langle \mathbf{a}^2(x - v_g t) \right\rangle\right)^{1/2}, \qquad (4.1)$$

where $\mathbf{a}^2(x - v_g t)$ is the dimensionless amplitude of the laser pulse. The resulting phase velocity $v_p = v_g$ is thus close to c as desired. Actually, the longitudinal PF of a laser pulse has two spikes with opposite sign on the rising and falling fronts, so that the pulse action can not be modeled as a single delta-like unipolar pulse as in Sect. 3.6.1. However, if the pulse duration τ_L is close to half the plasma oscillation period, $\tau_L \simeq T_p/2 = \pi/\omega_p$, the electron velocity reverses its sign at the time of the second PF kick, which will then enforce the oscillation. The situation is depicted in Fig. 4.1a. High-power laser pulses with suitably short durations (tens of femtoseconds for typical gas densities) were not available yet at the time of the paper by Tajima and Dawson (1979) who then suggested to use two laser beams of slightly different frequency and use the resulting beats to excite the wakefield. Such "beat wave" approach has now been almost abandoned after a few years of investigation.

In the linear approximation for the plasma wave and for a non-evolving laser pulse, i.e. fixing f_p in Eq. (4.1), the amplitude of the wakefield and its dependence on the pulse duration can be determined by the following calculation, using a Fourier Transform (FT) and a simple integration in the complex plane. We use the linearized fluid equations with the force pulse f_p as a driver,

$$\partial_t \delta n_e = -n_0 \partial_x u_x, \quad m_e \partial_t u_x = -e E_x + f_p - m_e \eta u_x, \quad \partial_x E_x = 4\pi e \delta n_e, \quad (4.2)$$

where we inserted a small "friction" term $-m_e \eta u_x$ to ensure causality, as will be evident below. We expand these equation in plane waves, i.e. we take the double FT according to the definition $\tilde{F}(k, \omega) = \int \int F(x, t) e^{-ikx + i\omega t} dx dt$ Because of (4.1), we obtain $\tilde{f}_p(k, \omega) = 2\pi \hat{f}_p(k) \delta(\omega - k v_g)$ where $\hat{f}_p(k) = \int f(x') e^{ikx'} dx'$ is the

---

[1] Here we do not consider so-called plasma wakefield accelerators where the wake wave is driven by beams of either electrons (Blumenfeld et al. 2007) or protons (Caldwell et al. 2009) generated by conventional accelerators.

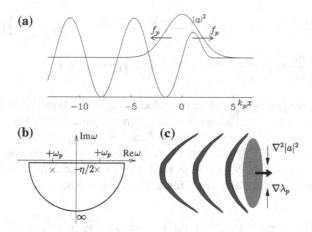

Fig. 4.1 **a** Excitation of a wakefield by a short laser pulse. The double ponderomotive kick on the rise and falling edges of the pulse enforces the oscillations. **b** Contour path in the complex plane to evaluate the integral (4.4). **c** Transverse structure of the wakefield excited by a laser pulse of finite width, showing the curved phase fronts

FT of the force pulse profile. Notice that $\hat{f}_p(k) = \hat{f}_p^*(-k)$ since f_p is real. We thus obtain by simple algebra the FT of the density perturbation

$$\delta\tilde{n}_e(k, \omega) = 2\pi \frac{n_0}{m_e} \hat{f}_p(k)\, \delta(\omega - kv_g) \frac{i\omega}{\omega^2 - \omega_p^2 + i\eta\omega}, \tag{4.3}$$

and by applying the inverse FT

$$\delta n_e(x, t) = \frac{n_0}{m_e} \int e^{-i\omega(t - x/v_g)} \frac{i\omega}{\omega^2 - \omega_p^2 + i\eta\omega} \hat{f}_p\left(\frac{\omega}{v_g}\right) d\omega. \tag{4.4}$$

The integral can be evaluated using Jordan's lemma. We go in the complex ω plane and we integrate over a path including the real axis and closing itself on a circle or arbitrary radius, extending either in the upper $(\mathrm{Im}(\omega) > 0)$ or lower $(\mathrm{Im}(\omega) < 0)$ plane depending on the sign of the argument of the exponential function, i.e. if $x - v_g t > 0$ or $x - v_g t < 0$. The result is the sum of residues of the integrand at the poles inside the path. In the complex plane, we find two poles below the real axis at $\omega = \pm\omega_p - i\eta/2$ (Fig. 4.1b). Then, the integral vanishes if $x - v_g t > 0$ since the path in the upper complex plane does not include any pole. Having established that $\delta n_e \neq 0$ only if $x - v_g t < 0$, we can let now $\eta = 0$ and write the result of the integral as $2\pi i$ times the sum of the residues at $\omega = \pm\omega_p$, obtaining

$$\delta n_e(x,t) = \frac{n_0}{m_e}\frac{1}{2}\left[e^{-i\omega_p(t-x/\upsilon_g)}\hat{f}_p\left(\frac{+\omega_p}{\upsilon_g}\right) + e^{+i\omega_p(t-x/\upsilon_g)}\hat{f}_p\left(\frac{-\omega_p}{\upsilon_g}\right)\right]\Theta\left(\upsilon_g t - x\right)$$

$$= -\frac{n_0}{m_e}\mathrm{Re}\left[e^{-i\omega_p t + ikx(\omega_p/\upsilon_g)}\hat{f}_p\left(\frac{\omega_p}{\upsilon_g}\right)\right]\Theta\left(\upsilon_g t - x\right). \qquad (4.5)$$

The wake has frequency ω_p and phase velocity $\upsilon_p = \upsilon_g$. Equation (4.5) allows to obtain the explicit form of the linear wake given the pulse shape. The amplitude is proportional to the spectral component of f_p at $k = \omega_p/\upsilon_g$, which is maximized if the length of the pulse τ_p is $\tau_p = \pi/kc = T_p\upsilon_g/2c$. As an example assume a square profile for γ_a in (4.1), i.e. $\gamma_a(x) = \gamma_0$ for $|x| < c\tau_p/2$ and zero elsewhere, which corresponds to a "double delta-kick" force on electrons. Then

$$\hat{f}_p(k) = -m_ec^2\int e^{-ikx}\partial_x\gamma_a dx = -ikm_ec^2\gamma_0\int_{-c\tau_p/2}^{+c\tau_p/2}e^{-ikx}dx$$

$$= -2m_ec^2\gamma_0 k\sin\left(\frac{kc\tau_p}{2}\right). \qquad (4.6)$$

Thus the wake amplitude is proportional to $\sin\left(c\omega_p\tau_p/2\upsilon_g\right)$ that is maximized for $\tau_p = (\pi\upsilon_g/\omega_pc)$ and vanishes for $\tau_p = (2\pi\upsilon_g/\omega_pc)$, as the two kicks compensate each other.

In the ultrarelativistic limit in which the amplitude of the driving laser pulse $a_0 \gg 1$ and the wake wave is highly nonlinear, the wavelength λ_p of the wake wave depends on a_0. To estimate such dependence, we observe that the electric field amplitude E should be such to balance the longitudinal PF. From $f_p = -m_ec^2\partial_x\left(1 + \langle a^2\rangle\right)^{1/2}$ and assuming $a_0 \gg 1$ we estimate the peak amplitude of the $f_{p0} \simeq m_eca_0/\tau_p$ with $\tau_p \simeq \pi/\omega_p$ (since $\upsilon_g \simeq 1$). If the wakefield is driven close to the wavebreaking limit (3.70), then $eE \simeq m_e\omega_pc\sqrt{2(\gamma_p - 1)} \simeq m_e\omega_p^2\lambda_p/4$ where (3.71) has been used. Posing $eE := f_{p0}$ we eventually obtain $\lambda_p \simeq (4/\pi)(c/\omega_p)a_0$. A rigorous self-consistent calculation (see e.g. Esarey et al. 1996; Bulanov et al. 2001) gives $\lambda_p \simeq 2^{3/2}(c/\omega_p)a_0$ for $a_0 \gg 1$.

The dependence of λ_p on the laser amplitude has important consequences for the transverse structure of the wakefield generated by a laser pulse of finite width: as the intensity $|a^2(r)|$ typically decreases with the distance r from the laser axis, the local wavelength $\lambda_p(r)$ will also decrease and the phase fronts will therefore be curved as sketched in Fig. 4.1c. A density depression along the axis, that may be created by the transverse ponderomotive force, has a similar effect since the local plasma frequency $\omega_p(r)$ would decrease with r and, for fixed $\upsilon_p = \upsilon_g$, the wavelength $\lambda_p(r) = 2\pi\upsilon_g/\omega_p(r)$ would increase. Such curved wavefronts have been experimentally observed using an ultrafast imaging technique (Matlis et al. 2006). The transverse structure of the wakefield can lower the wavebreaking threshold, because the resulting electron oscillations may self-intersect more easily than in plane geometry (Bulanov et al. 1997), and also plays a crucial role in the concept of relativistic flying mirrors (Sect. 6.2).

4.1.2 Wakefield Acceleration

In the original paper by Tajima and Dawson (1979), an estimate for the maximum energy gain for an electron in a wakefield $E_x = E_0 \cos(k_p x - \omega_p t)$ is obtained as follows. First, $\omega \ll \omega_p$ and thus $v_p = \omega_p / k_p = v_g \simeq c$ are assumed, while the non-relativistic wave-breaking limit $E_0 = m_e \omega_p c / e$ is taken as an upper limit of the field, although the latter assumption is questionable for a high-amplitude wave with $v_p \simeq c$ (Sect. 3.6.2). For the sake of simplicity is however convenient to deal with an harmonic, linear wave. To evaluate the energy gain of an electron we make a Lorentz transformation from the laboratory frame S to the frame S' moving with velocity v_p, where the wake has wavevector and frequency

$$k'_p = \gamma_p (k_p - v_p \omega_p / c^2) = \omega_p / (v_p \gamma_p), \qquad \omega'_p = \gamma_p (\omega_p - v_p k_p) = 0, \quad (4.7)$$

so that the electric field is static in S'. Notice that $\gamma_p = (1 - v_p^2/c^2)^{1/2} = \omega/\omega_p$. The longitudinal field E_x is invariant, so that the electric field of the wake in S' is of the form $E'_x = E_0 \cos(k'_p x')$, corresponding to a potential $\Phi' = -(E_0/k'_p)\sin(k'_p x')$. In such potential, an electron will make the highest energy gain $W' = 2eE_0/k'_p \simeq 2m_e c^2 \gamma_p = 2m_e c^2 \omega/\omega_p$ when starting at rest from a maximum of the potential energy $-e\Phi'$ and descending down to a minimum. The energy-momentum in S' is thus $(\mathcal{E}', p'_x c) \simeq (W', W')$ and transforming back in S the electron energy is

$$\mathcal{E} = \gamma_p (\mathcal{E}' + v_p p'_x) \simeq 2\gamma_p W' = 2m_e c^2 (\omega/\omega_p)^2, \qquad (4.8)$$

that almost corresponds to the energy gain W. Notice that in estimating W' we considered the "luckiest" electron which would enter the wave with an initial velocity $\sim v_p$ at a maximum of the potential energy and leave the wave immediately after having completed half an oscillation in the wakefield. In the laboratory frame, we would see such electron get trapped in the wave at some point and leave it after an accelerating distance L_{acc} that is easily estimated thanks to the invariance of E_0:

$$L_{acc} = \frac{W}{eE_0} \simeq \frac{2\omega^2 c}{\omega_p^3} = \frac{\lambda}{\pi}\left(\frac{\omega}{\omega_p}\right)^3, \qquad (4.9)$$

corresponding to an accelerating gradient $\mathcal{E}/L_{acc} = (2m_e c^2/\lambda)(\omega_p/\omega)$. Assuming a 100 MeV (10 GeV) accelerator with $\omega/\omega_p = 10\,(100)$ according to (4.8) and corresponding to $n_e = 10^{19}$ cm^{-3} (10^{17} cm^{-3}) for $\lambda = 1\,\mu$m, then $L_{acc} \simeq 300\,\mu$m (3 mm), which is impressively small with respect to the size of "standard" accelerators. However, the point is to generate a plasma wake as long as L_{acc}, which is possibly much larger than the Rayleigh length over which the driving laser pulse would diffract. Thus, one of the challenges of the development of laser-plasma electron accelerator has been the guiding of the laser pulse over long distances, using several approaches such as, e.g. searching for a stable self-guiding regime near the self-focusing threshold (see Sect. 3.3) or preforming a low-density channel in the plasma.

Besides obtaining longer acceleration lengths and higher energies, the research performed over about three decades has aimed at improving the electron beam quality, including key features such as narrow energy spread and low divergence and emittance. The related experimental and theoretical work has been extensively reviewed by Esarey et al. (2009), thus here we only briefly summarize major issues and results.

Two major approaches to the development of an efficient laser plasma accelerator might be characterized by the strategy to give the electrons an optimal phase for subsequent acceleration. The first approach is based on self-injection, i.e. on finding a regime in which self-trapping of electrons into the wakefield occurs with the desired phase. As seen in Sect. 3.6.2 a plasma wave should "break" when the oscillation velocity $u_x = v_p$, that looks identical to the optimal phase condition. Thus, driving the plasma wave near the breaking threshold should increase both the energy and number of electrons self-trapped in the wave, as it was demonstrated experimentally in 1995 for a relatively long pulse experiment (Modena et al. 1995).[2] Later numerical experiments (Mora and Antonsen 1996; Pukhov and Meyer-ter-Vehn 2002) showed that ultrashort and superintense laser pulses may drive the electron density perturbation so strongly that a cavity forms behind the laser pulse. Electrons are trapped inside the cavity and accelerated to high energy with a narrow spectrum, while in the tail of the moving cavity the plasma wave is broken, so that no relevant acceleration occurs there. In addition, the laser pulse is self-guided over distances exceeding the diffraction length. This scenario has become popular as the "bubble regime". Two experiments performed in this regime and reporting monoenergetic electron bunches (Mangles et al. 2004; Faure et al. 2004) were published in a celebrated "dream beam" issue of the Nature journal in 2004. A third experiment reported in the same issue also used self-injection but employed a preformed plasma channel to increase the acceleration length up to the dephasing limit (Geddes et al. 2004). This latter approach has later demonstrated monoenergetic acceleration up to the GeV limit (Hafz et al. 2008), similarly to what has been obtained by using a capillary discharge as a waveguide (Leemans et al. 2006). The improvement of electron beam stability and emittance by the use of an external magnetic field able to sustain the preformed channel has been also reported (Hosokai et al. 2010). An alternative approach is based on generating the plasma wake in a decreasing density profile, for which v_p decreases along the propagation path eventually reaching the breaking threshold (Bulanov et al. 1998); this scheme has also been experimentally investigated (Gonsalves et al. 2011, and references therein).

The second approach considers the stimulated injection of electrons to be accelerated with the desired energy and quality. The extremely short spatial and tempo-

---

[2] In experiments where the $\tau_l > T_p$, the driving pulse and the plasma wave overlap and nonlinear beats generate EM sidebands at $\omega \pm n\omega_p$ ($n = 1, 2, \ldots$), a process known as Raman Scattering. In a strongly nonlinear regime, the density perturbation of the plasma wave self-modulates the envelope of the laser pulse resulting in a positive feedback for growth. See Esarey et al. (2009) for details on this regime.

ral scales associated to the wakefield dictated the development of optical injection techniques using laser pulses shorter than the plasma period, like the pump pulse generating the wakefield. In a nutshell, the idea is that one (or more) laser pulses intersecting the wakefield with a proper phase may locally accelerate ponderomotively some electrons up to the optimal velocity for being trapped in the plasma wave. The optical injection approach has the promise of a better external control upon the acceleration process at the expense of a more complex set-up. Using an injection pulse counterpropagating to the pump one, an improved stability of the acceleration has been achieved (Faure et al. 2006).

4.2 Overdense Plasmas

The generation of high energy electrons, commonly named *fast* or *hot* electrons, is also observed in high-intensity interaction with overdense plasmas, which typically corresponds to the use of solid targets in experiments. A major part of such electrons penetrates beyond the cut-off layer and is ordinarily the most important mechanism of energy transport inside the target. Several experimental observations and numerical simulations suggest that, for a laser pulse of dimensionless amplitude a_0, the typical order of magnitude of the fast electron energy is given by

$$\mathcal{E}_p = m_e c^2 \left(\sqrt{1 + a_0^2/2} - 1 \right), \tag{4.10}$$

that is also called the "ponderomotive" energy. This relation and its name may be roughly justified by observing that the reflection of the laser pulse from the overdense plasma generates a standing wave, e.g. $\mathbf{a}(x, t) \simeq \mathbf{a}(x) \cos \omega t$ and thus a ponderomotive potential $\Phi_p \simeq m_e c^2 \left(1 + \mathbf{a}^2(x)/2\right)^{1/2}$ (Sect. 2.1.4). However, the process of fast electron generation is certainly far from being as simple as the motion of a particle in a potential, and possibly not completely understood yet. A satisfactory theory should explain several observations beyond the scaling (4.10), such as the pulsed nature of fast electron generation, the conversion efficiency of the laser energy, the dependence on the plasma parameters, and the energy spectra of electrons.

It should also be mentioned that detailed experimental characterization of fast electron production is a difficult task for several reasons, including the lack of a complete control of the interaction conditions. As an example, short pulse laser systems have been affected unavoidably for a long time by prepulses and long-duration pedestals preceding the main pulse and intense enough to cause early plasma formation, modifying the effective density and scalelength of the target. These issues made it historically difficult to perform ultimate benchmarks of theoretical models, but most recent developments in laser technology should progressively lead to more controlled experiments. A complete account of research on this topic is beyond our scope. In the following we limit ourselves to outline some basic facts and to describe related simple models.

4.2.1 Routes to Collisionless Absorption

We begin by briefly describing some possible mechanisms of collisionless absorption of laser energy in overdense plasmas, with the main aim to clarify the conditions in which the generation of fast electrons directed into the target may occur or not, and the assumptions on which the models of Sects. 4.2.2 and 4.2.3 are based.

Let us consider an ultrashort, high-intensity pulse impinging on a solid target. We assume that ionization and heating of the solid material occur rapidly enough so that the target can be considered as a plasma with step-like ion density profile $n_i = n_0 \Theta(x)$ as long as hydrodynamic expansion is negligible. Due to the high electron density ($n_0 \gg n_c$) in a solid material the EM field will be evanescent inside the plasma according to (3.5) penetrating only in the "skin" layer of thickness ℓ_s defined by (3.8). Notice that if $\varepsilon(\omega)$ is real, then total reflection and no energy absorption occur. Absorption requires $\varepsilon(\omega)$ to have an imaginary part, which is indeed the case if in the calculation leading to (3.5) we add a friction force $-m_e \nu_c \mathbf{u}_e$ to the equation of motion of electrons, obtaining

$$\varepsilon(\omega) = 1 - \frac{\omega_p^2}{\omega(\omega + i\nu_c)}. \tag{4.11}$$

The latter expression leads, for $(\omega_p/\omega)^2 = n_0/n_c \gg 1$, to an absorption coefficient $A \simeq 2\nu_c \omega^2/\omega_p^3$ as obtained from Fresnel formulas (Jackson 1998, Sect. 7.3). In a classical plasma the friction is determined by collisions with ions, so that we talk of collisional absorption (also named as *inverse Bremsstrahlung*, see Mulser and Bauer 2010, Sect. 3.4). For a non-relativistic electron of velocity υ_e and energy $\mathcal{E}_e = m_e \upsilon_e^2/2$ the collision rate $\nu_c = n_i \sigma_c \upsilon_e$ where $\sigma_c \propto \upsilon_e^{-4} \propto \mathcal{E}_e^{-2}$ is the Coulomb cross section. Thus, as heating proceeds and the average value of $\mathcal{E}_e$ increases, collisional absorption becomes inefficient because of the scaling $\nu_c \sim \mathcal{E}_e^{-3/2}$ (*runaway effect*). In high-intensity interactions the rise in $\mathcal{E}_e$ is due to the coherent motion in the laser field, rather than to an increase in thermal energy (i.e. in temperature), so that collisional absorption decreases with growing intensity. Thus, efficient energy absorption at high intensity may only occur via collisionless mechanisms.

To briefly and heuristically introduce a class of kinetic mechanisms leading to collisionless absorption let us look closer at the skin layer of the plasma, see Fig. 4.2. We assume that due to ionization and collisional absorption the electrons have been already heated up to a temperature $T_e = m_e \upsilon_{te}^2/2$, with υ_{te} the thermal velocity. Because of their thermal energy electrons attempt to escape on the vacuum side ($x < 0$) but they are electrostatically confined: on the average, a sheath region, i.e. a thin charge separation layer of thickness ℓ_c will be produced with a corresponding

Fig. 4.2 Sketch of field and density profiles in the skin layer of a solid-density plasma. The sheath field E_x "reflects" electrons from the bulk

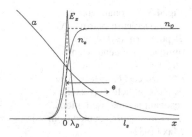

field E_x. Since E_x must backhold electrons of energy T_e, we estimate $eE_x\ell_c = T_e$, and using a dimensional argument, since n_0 and T_e are the only parameters from which we may obtain a length, we have $\ell_c \simeq \lambda_D = v_{te}/\omega_p$. Thus, looking on a spatial scale $\sim\ell_s \gg \lambda_D$ and on a time scale $\sim\omega^{-1} \gg \omega_p^{-1}$ we may consider electrons as instantaneously "reflected" from the sheath field as a reflecting boundary at $x = 0$. Such reflection plays the part of an effective collision, with frequency $v_c \sim \ell_s/v_{te} = \omega_p v_{te}/c$ leading to a non-vanishing absorption, so that this scenario is named as *sheath inverse Bremsstrahlung* (SIB) (Catto and More 1977). If, in addition to the onset of sheath reflection, the time $\tau_s \simeq v_c^{-1} = \ell_s/v_{te}$ in which electrons cross the skin layer is shorter than the period of the EM field, i.e. $\tau_s \lesssim 2\pi/\omega$, the oscillation velocity may not average to zero over a period because the electrons will have already moved into a region of weak field. This leads to increased absorption and also to deeper penetration of the EM wave in the plasma, so that the skin depth changes and must be determined self-consistently. This is a very simplified description of the *anomalous skin effect* (ASE) (Weibel 1967). A rigorous theory of ASE requires a solution of the Vlasov-Maxwell system for a sharp edged plasma where the sheath reflection enters as a boundary condition. In appropriate regimes, ASE absorption plus collisional energy transport inside the dense and cold target gives a fair model to explain short-pulse absorption measurements (Price et al. 1995; Rozmus et al. 1996). However, ASE (as well as SIB) fails to explain the origin of "fast", highly suprathermal electrons with energies close to (4.10) and largely exceeding the "bulk" value T_e, just because the EM field in the skin layer is much less than the vacuum field. In addition, the sheath reflection condition is consistent only if the external field is too weak to drag electrons in vacuum against the sheath potential, thus the theory is limited to low intensities.

Another general possible route to absorption and field enhancement in an overdense plasma is based on the excitation of resonances, i.e. of normal modes of the plasma. In an overdense plasma the basic *resonance absorption* mechanism consists in the excitation of plasma oscillations in the region where $\omega = \omega_p$, i.e. at the critical surface. As a necessary condition the EM wave must have oblique incidence and P-polarization. The situation is illustrated in Fig. 4.3: only if the electric field is parallel to the density gradient a charge density perturbation may be produced, exciting a plasma oscillation.

To elucidate this point further let us consider the following simple electrostatic or "capacitor" model, where the EM wave is described via an external, oscillating field $\mathbf{E}_d = \mathrm{Re}\left(\tilde{\mathbf{E}}_d e^{-i\omega t}\right)$ and the plasma is inhomogeneous with background density

Fig. 4.3 Resonant absorption
in an overdense plasma. A
P-polarized EM wave
obliquely incident at an
angle θ is reflected at the
$n_e = n_c \cos^2 \theta$ surface. The
evanescent field may excite a
resonant plasma oscillation at
$n_e = n_c$. Figure adapted from
Max (1982)

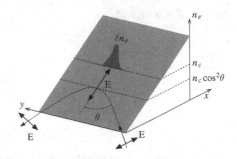

$n_0 = n_0(x)$. The equations describing an electrostatic perturbation are

$$\mathbf{\nabla} \cdot \mathbf{E} = -4\pi e(n_e - n_0), \quad \partial_t n_e = -\mathbf{\nabla} \cdot (n_e \mathbf{u}), \quad \frac{d\mathbf{u}}{dt} = -\frac{e}{m_e}(\mathbf{E} + \mathbf{E}_d), \quad (4.12)$$

where, as usual, $d/dt = \partial_t - \mathbf{u} \cdot \mathbf{\nabla}$. We linearize the equations in the limit $u_x \ll L/\omega$
where $L = n_0/|\mathbf{\nabla} n_0|$ is the density scalelength. This approximation corresponds to
take the density as uniform over the electron oscillation amplitude. For the density
perturbation $\delta n_e = n_e - n_0 = \mathrm{Re}\left(\delta \tilde{n}_e e^{-i\omega t}\right)$ we thus obtain $-i\omega\left(\delta \tilde{n}_e\right) \simeq -n_0 \mathbf{\nabla} \cdot$
$\tilde{\mathbf{u}} - \tilde{\mathbf{u}} \cdot \mathbf{\nabla} n_0$ and, by eliminating $\mathbf{u}$,

$$\delta n_e = \frac{1}{4\pi e} \frac{(\mathbf{E} + \mathbf{E}_d) \cdot \mathbf{\nabla} n_0}{n_0(x) - n_c}. \quad (4.13)$$

The appearance of a singular point at $n_0(x) = n_c$ shows the onset of the resonance
at the "critical" surface, where $\omega = \omega_p(x)$. However, a necessary condition for the
excitation of such resonance is that the source term $\mathbf{E}_d \cdot \mathbf{\nabla} n_0 \neq 0$, i.e. the driving
field must have a nonzero component along the density gradient. Simple geometric
considerations show that if $\mathbf{E}_d$ represents the field of an impinging EM wave, the
latter must be obliquely incident and P-polarized, as sketched in Fig. 4.3. This issue
complicates the picture because at oblique incidence the EM wave is reflected at
densities lower than n_c. In fact, for a plane wave the wavevector component along
the surface, $k_y = (\omega/c)\sin\theta$, must be conserved across the reflection layer. Putting
$k^2 = k_x^2 + k_y^2$ in the dispersion relation (3.6) we get $\omega^2 = \omega_p^2 + k_x^2 c^2 + \omega^2 \sin^2\theta$ and
thus $k_x^2 < 0$ when $n_e > n_c \cos^2\theta$. The EM field will be evanescent in this region
and thus, depending on the value of L, it might be too weak at $n_e = n_c$ to excite
the resonance. It turns out that an optimal angle exist for given L, as a compromise
between maximizing the electric field component perpendicular to the plane and its
amplitude. A satisfactory modeling of resonant absorption is not trivial and may be
found in the literature (see e.g. Mulser and Bauer 2010, Chap. 4).

   If the temperature is finite, plasma oscillations acquire a group velocity (see
Sect. 3.1) and propagate as ES waves in the plasma according to the dispersion
relation (3.12), i.e. "downhill" the density profile in the region where $\omega_p < \omega$ and k
is real. Such longitudinal waves may accelerate a fraction of thermal electrons along

the propagation direction, i.e. out of the target surface. To consider electron acceleration in the *forward direction* leading to fast electrons entering the target bulk, one must go beyond the simple picture of linear resonance absorption as described in the next Sect. (4.2.2).

A very steep density gradient smears the plasma resonance out but also leads to the appearance of another normal mode: electron *surface waves* (SWs) or *surface plasmons*, which are localized along the surface and propagate along the latter with dispersion relation (Landau et al. 1984)

$$k^2 c^2 = \omega^2 \frac{\varepsilon(\omega)}{1 + \varepsilon(\omega)} = \frac{1 - \omega_p^2/\omega^2}{2 - \omega_p^2/\omega^2}, \tag{4.14}$$

so that the waves exist for $1 < \omega_p/\omega < \sqrt{2}$. The components of the electric field, for propagation at the $x = 0$ surface and along the y axis, may be written as

$$E_x = ik E_0 \left[\Theta(-x) \frac{e^{+q_< x}}{q_<} - \Theta(+x) \frac{e^{-q_> x}}{q_>} \right] e^{-i\omega t},$$
$$E_y = E_0 \left[\Theta(-x) e^{+q_< x} + \Theta(+x) e^{-q_> x} \right] e^{-i\omega t}, \tag{4.15}$$

where $q_> = (\omega/c)(\omega_p^2/\omega^2 - 1)(\omega_p^2/\omega^2 - 1)^{-1/2}$ and $q_< = (\omega/c)(\omega_p^2/\omega^2 - 1)^{-1/2}$.

Resonant excitation of SWs is a candidate for field enhancement and also for electron acceleration along the surface itself, having the SW electric field a component along $\mathbf{k}$. The difficulty is that (4.14) implies a phase velocity $v_p < c$, so that a phase matching with an incident EM wave is not possible. The way out of this difficulty is to use a "grating" target with a periodically modulated surface with wavevector k_g. In fact, now (in the limit of plane waves) the medium can be considered as periodic, so as a consequence of Floquet's theorem (equivalent to Bloch's theorem in solid state physics) solutions of the wave equation are of the form $\sim e^{iky} f(y)$ with $f(y) = f(y + 2\pi/k_g)$. Phase matching of Floquet-Bloch SWs (whose dispersion relation is affected only slightly by the presence of the grating) with an external field of wavevector component $k_y = (\omega/c) \sin\theta$ requires a condition $k = k_y + n k_g$, with n an integer number, that can be satisfied at a proper angle for given k_g. Experimental and theoretical studies of SW absorption in laser-plasma interactions may be found in the recent literature (see e.g. Bigongiari et al. 2011 and references therein). Here we only remark, for the reader's curiosity, that the absorption model of next section proposed by Brunel (1987), which has become very popular in laser-plasma interaction physics, was originally motivated as a study of damage threshold in a SW-based electron accelerator concept.[3]

---

[3] We also may notice that surface plasmons, both propagating and spatially localized, and their coupling with external waves in structured targets are building blocks of *plasmonics* and its several applications (see e.g. Barnes et al. 2003; Ozbay 2006 and references therein) and we might speculate that these concepts would find other applications with intense laser pulses, in what we might call high-field plasmonics.

4.2.2 "Not-So-Resonant" Absorption and "Vacuum Heating"

The above presented model of resonant absorption is valid in the limit $u_x/\omega \ll L$ that implies the density to be taken as nearly uniform over the oscillation amplitude. If this condition is violated, a local plasma frequency cannot be defined. If, furthermore, the field is so intense that $u_x/\omega > c/\omega_p$, we expect plasma oscillations to break and cause electron heating. These conditions are favored by intense fields and sharp density gradients, and are the basis for the electrostatic model of "not-so-resonant absorption" proposed by Brunel (1987) and also referred to either as "Brunel effect" or "vacuum heating"(see Gibbon 2005, p. 161) in the literature. In the model, the gradients of both the plasma density and the driving field are assumed to be infinite, as a step-like plasma profile with an external capacitor field extending on the vacuum side only are assumed. Electrons are dragged out in vacuum for about half a cycle and then re-enter the plasma side there delivering their energy. The model thus accounts in a simplified way for the pulsed generation (once per laser cycle) of fast electrons directed into the target and having an energy, roughly speaking, close to the "vacuum" value.

To drag electrons out of the plasma, the external driver must suppress the sheath field barrier (due to E_x in Fig. 4.2). For $\omega_p \gg \omega$ and nearly total reflection, the laser field component normal to the surface has an amplitude $E_\perp \simeq 2E_0 \sin\theta$, so that posing $E_\perp > T_e/(e\lambda_D) = (4\pi n_0 T_e)^{1/2}$ leads to a condition $4(I/c)\sin^2\theta > n_0 T_e$ with $I = cE_0^2/4\pi$ the laser intensity. Hence, at least for not very large angles we infer that the radiation pressure $2(I/c)\cos^2\theta$ typically exceeds the plasma pressure $n_0 T_e$, that implies that the ponderomotive force is strong enough to counteract the thermal expansion and to steepen the density profile, making the assumption of a step-like plasma more self-consistent. However, this rough statement neglects both the strong evanescence of the EM field in the skin layer and its more general dependence on θ and n_0/n_c.

To provide a "minimal" analytical description of the dynamics of electron acceleration at a steep interface we again use an electrostatic, capacitor model (Mulser et al. 2001) slightly refined with respect to Brunel (1987). With respect to the model for resonance absorption in Sect. 4.2.1 we now keep nonlinear terms and assume a step-boundary plasma with ion density profile $n_i = n_0\Theta(x)$ ($Z = 1$ for simplicity) and 1D geometry. The total electric field is the sum of the electrostatic and the driver fields, e.g. $E_x = E_{es} + E_d$ where $E_d = E_d(x,t) = \tilde{E}_d(x,t)\sin\omega t$ as a possible form. The equations are thus

$$\partial_t n_e = -\partial_x(n_e u_x), \quad \partial_x E_{es} = 4\pi e(n_i - n_e), \quad \frac{du_x}{dt} = -\frac{e}{m_e}(E_{es} + E_d). \quad (4.16)$$

It is convenient to switch to *Lagrangian* variables x_0 and τ defined by

$$x = x_0 + \xi(x_0,\tau), \quad \tau = t, \quad d\xi/dt = u_x. \quad (4.17)$$

The physical meaning is the following: x_0 is the initial coordinate of an electron (more precisely, of an individual fluid element) and $\xi = \xi(x_0, \tau)$ is the displacement from the initial position at the time $\tau = t$. Notice that for 1D motion each "electron" is actually a negative charge sheet. The transformation rules for the variables give

$$\partial_0 = \partial_0 x \partial_x + \partial_0 t \partial_t = (1 + \partial_0 \xi)\partial_x, \quad \partial_\tau = \partial_\tau x \partial_x + \partial_\tau t \partial_t = u_x \partial_x + \partial_t. \quad (4.18)$$

The second equation gives $\partial_\tau = d_t$, so that the equation of motion is now *linear*. The first equation shows that the transformation to Lagrange variables is singular when $(1 + \partial_0 \xi) = 0$. In fact, two trajectories originating from infinitesimally close initial positions x_0 and $x_0 + dx_0$ will intersect at the time τ if

$$x_0 + \xi(x_0, \tau) = x_0 + dx_0 + \xi(x_0 + dx_0, \tau), \quad (4.19)$$

from which a Taylor expansion $\xi(x_0 + dx_0, \tau) = \xi(x_0, \tau) + \partial_0 \xi(x_0, \tau)dx_0$ immediately brings the condition $(1 + \partial_0 \xi) = 0$. The singularity is also apparent in the solution for n_e. The continuity equation becomes

$$0 = \partial_t n_e + \partial_x (n_e u_x) = \partial_t n_e + u_x \partial_x n_e + n_e \partial_x u_x = \partial_\tau n_e + n_e \frac{\partial_0 (\partial_\tau \xi)}{1 + \partial_0 \xi}, \quad (4.20)$$

that can be rearranged as $\partial_\tau n_e / n_e = -\partial_\tau (\partial_0 \xi / (1 + \partial_0 \xi))$. Integration of this latter equation brings

$$n_e = n_e(x_0, \tau) = \frac{n_e(x_0, 0)}{1 + \partial_0 \xi(x_0, \tau)} = \frac{n_0 \Theta(x_0)}{1 + \partial_0 \xi(x_0, \tau)}, \quad (4.21)$$

which is infinite when $1 + \partial_0 \xi = 0$, marking the breaking of the fluid description.
Poisson's equation for E_{es} becomes

$$\frac{\partial_0 E_{es}}{1 + \partial_0 \xi} = 4\pi e n_0 \left(\Theta(x_0 + \xi) - \frac{\Theta(x_0)}{1 + \partial_0 \xi} \right), \quad (4.22)$$

with the following solution (continuous at $x = 0$)

$$E_{es} = \begin{cases} +4\pi e \, n_0 \xi & (x_0 + \xi > 0) \\ -4\pi e \, n_0 x_0 & (x_0 + \xi < 0) \end{cases}. \quad (4.23)$$

The equations of motion describing electrostatic, forced oscillations of plasma sheets across a step-like interface are then

$$\partial_\tau^2 \xi = \begin{cases} -\omega_p^2 \xi - e E_d / m_e & (x_0 + \xi > 0) \\ +\omega_p^2 x_0 - e E_d / m_e & (x_0 + \xi < 0) \end{cases}. \quad (4.24)$$

Electrons for which $x_0 + \xi > 0$ at any time never cross the interface at $x = 0$; their equation of motion is that of a simple forced harmonic oscillator of natural frequency ω_p, which is resonantly driven if $\omega_0 = \omega_p$, as we already know. The situation is different for electrons crossing the boundary $(x = x_0 + \xi < 0)$; these feel a secular force $\omega_p^2 x_0$ leading to a net acceleration and to a dephasing between $u_x = \partial_\tau \xi$ and E_d, so that $\langle u_x E_d \rangle \neq 0$.

Following Mulser et al. (2001), we seek for a solution of (4.24) for a driver $E_d = \tilde{E}_d \Theta(t) \sin \omega t$ which is turned up abruptly at $t = 0$ (we write again t for τ in the following). Since (4.24) is discontinuous we need to match the "inner" $(x > 0)$ and "outer" $(x < 0)$ solutions at the time $t = t_0$ when an electron starting at x_0 crosses the $x = 0$ interface. For $x = x_0 + \xi > 0$, the solution is straightforward

$$\xi = -\frac{e\tilde{E}_e}{m_e(\omega_p^2 - \omega^2)} \sin \omega t = -\frac{u_d}{\omega}\frac{\sin \omega t}{\omega_p^2/\omega^2 - 1}, \qquad (x_0 + \xi < 0), \qquad (4.25)$$

where we put $u_d \equiv e\tilde{E}_e/(m_e/\omega)$. The electron crosses the $x = 0$ boundary at the time t_0 given by the condition

$$\xi(t_0) = -\frac{u_d}{\omega}\frac{\sin \omega t_0}{\omega_p^2/\omega^2 - 1} := -x_0, \qquad (4.26)$$

with velocity $u_x(t_0) = -u_d(\omega_p^2/\omega^2 - 1)^{-1} \cos \omega t_0$.

For $x = x_0 + \xi < 0$ the solution satisfying $\xi(t_0) = -x_0$ and $(\partial_t \xi)(t_0) = u_x(t_0)$ is

$$\xi = -x_0 + u_d(\sin \omega t - \sin \omega t_0) - \frac{u_d \cos \omega t_0}{1 - (\omega/\omega_p)^2}(t - t_0)$$

$$+ \frac{u_d \omega}{2}(t - t_0)^2 \sin \omega t_0 \qquad (x_0 + \xi < 0). \qquad (4.27)$$

The last term shows the existence of a secular acceleration on the electrons, which gains a net velocity of the order of u_d. Brunel (1987) obtained a simplified Eq. (4.27) in the limit $\omega_p/\omega \to \infty$, that corresponds to neglecting the collective response of the plasma. From (4.27) one can find the velocity at which each electron re-enters the plasma region $(x < 0)$ at the time when $x = 0$, i.e. $\xi = -x_0$ again, and integrating over all the re-entering electrons the absorbed energy may be estimated. This procedure is cumbersome and possibly plagued by the onset of singularities in the Lagrange variables. However, one can easily solve Eq. (4.24) numerically for a discrete but large set of values of $x_0 > 0$, using Dawson's sheet model (Sect. 2.3.1) for a semi-infinite plasma. In this way, representative trajectories of electrons moving across the interface are found as in Fig. 4.4. Such trajectories include electrons which re-enter at high velocity inside the plasma, very similarly to what is observed in electromagnetic and self-consistent simulations (Sect. 4.2.4).

Still within Brunel's approximation, a connection between the capacitor model and the actual case of a P-polarized, obliquely incident wave might be made as

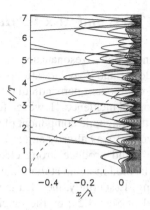

Fig. 4.4 Numerical solution of the sheet model (Sect. 2.3.1) based on Eq. (4.24). The sheets are spaced by $\simeq \lambda/5000$ and 1/30 of the trajectories in the (x, t) plane are shown. The initial density $n_0 = 5n_c$. The driver field has peak amplitude $0.8 m_e \omega c/e$ in vacuum, a $\sin^2(\pi t/2\tau)$ rising front with $\tau = 5T = 10\pi/\omega$ (*dashed line*), and has the spatial profile resembling a P-polarized laser field

follows. The driver field amplitude is taken equal to the electric field at the surface, $E_d \simeq 2E_0 \sin\theta$ in the limit $\omega_p \gg \omega$. From (4.24) the number of electrons (per unit surface) crossing the interface is $\simeq n_0 u_d/\omega_p^2 \simeq E_d/(4\pi e)$. Roughly assuming that each electron ejected on the vacuum side re-enters the plasma with velocity $\simeq u_d$ and within a period $2\pi/\omega$, the absorbed intensity is estimated as

$$I_{\text{abs}} \simeq \frac{(m_e u_d^2/2)(n_0 u_d/\omega)}{2\pi/\omega} = \frac{e E_d^3}{16\pi^2 m_e \omega} - \frac{e E_0^3 \sin^3\theta}{2\pi^2 m_e \omega}. \qquad (4.28)$$

Dividing by the incident energy flux $I_{\text{inc}} = (c/8\pi)E_0^2 \cos\theta$ one obtains an absorption coefficient $A \sim a_0 \sin^3\theta/\cos\theta$. The scaling with a_0 and θ has been fairly supported by experiments at low intensity for which it was possible to keep the $L \ll u_d/\omega$ condition (Grimes et al. 1999). Increasing either a_0 or θ, A grows non-physically above unity. To obtain a meaningful absorption coefficient $A \leq 1$, the energy depletion of the incident wave may be taken self-consistently into account by writing $E_d \simeq f(A)E_0 \sin\theta$ with $f(A) = 1 + \sqrt{1 - A}$. In addition, to account for relativistically strong intensities one replaces $m_e u_d^2/2$ with $m_e c^2 \left(\sqrt{1 + u_d^2} - 1 \right)$, eventually obtaining an implicit relation for $A = A(a_0, \theta)$:

$$A \simeq \frac{f(A)}{a_0} \left[(1 + f^2(A)a_0^2 \sin^2\theta)^{1/2} - 1 \right] \frac{\sin\theta}{\cos\theta}. \qquad (4.29)$$

The study of this expression shows A to peak at grazing angles for $a_0 \ll 1$ and to have a broad maximum at smaller angles for $a_0 \gg 1$, a limit in which A is independent on a_0 (see Gibbon 2005, Sect. 5.5.2).

4.2.3 Magnetic Force Effects and "J × B" Heating

Necessary conditions for Brunel's not-so resonant absorption are the same as resonant absorption, i.e. oblique incidence and P-polarization, because the driving force is provided by the electric field component perpendicular to the surface, that is absent both for S-polarization and normal incidence. However, for high intensities nonlinear oscillations can be driven by the magnetic component of the Lorentz force. Actually, two years before Brunel's work it was already suggested by Kruer and Estabrook (1985) that the $J \times B$ force could produce strong electron heating; although the mechanism was maybe not clarified in detail at that time, the effect has been known since then as "$J \times B$" heating. Essentially, the idea is to replace the driving field E_d in the electrostatic model of Sect. 4.2.2 by the longitudinal $v \times B$ force rather than the P-component of the electric field. One thus expects a similar dynamics, the main differences being that the driving force has a different dominant frequency (2ω instead of ω) and it scales as a_0^2, rather than a_0.

We go a bit deeper in the analysis of $v \times B$-force effects using a simple perturbative, non-relativistic approach. Still assuming a step-like density profile and considering the normal incidence of an elliptically polarized plane wave of amplitude a_0, in the linear approximation the vector potential inside the plasma can be written as

$$\mathbf{a}(x, t) = \frac{a(0)}{\sqrt{1 + \epsilon^2}} e^{-x/\ell_s} (\hat{\mathbf{y}} \cos \omega t + \epsilon \hat{\mathbf{z}} \sin \omega t), \qquad (4.30)$$

where ϵ is the ellipticity ($0 < \epsilon < 1$) and $a(0) = 2a_0/(1 + \mathsf{n})$ (from Fresnel formulas). Using (4.30) the $-e\mathbf{v} \times \mathbf{B}$ force can be written as

$$F_x = -m_e c^2 \partial_x \frac{\mathbf{a}^2}{2} = F_0 e^{-2x/\ell_s} \left(1 + \frac{1 - \epsilon^2}{1 + \epsilon^2} \cos 2\omega t \right), \qquad (4.31)$$

where $F_0 = 2m_e c^2 |a^2(0)|/\ell_s = (2m_e c^2/\ell_s)(\omega/\omega_p)^2 a_0^2$, being from (3.5) $|1 + \mathsf{n}|^2 = (\omega_p/\omega)^2$. The cycle average of (4.31) is the ponderomotive force, which is independent of the polarization. The second, oscillating term at the 2ω frequency vanishes for circular polarization ($\epsilon = 1$). This leads to a very different laser-plasma coupling between linear and circular polarization at normal incidence.

To proceed with our perturbative approach we now consider the longitudinal electron motion under the action of the force (4.31) and of the electrostatic field, using the linearized equations

$$m_e \partial_t u_x = -eE_x + F_x, \qquad \partial_t \delta n_e = -n_0 \partial_x u_x, \qquad \partial_x E_x = 4\pi e \delta n_e. \qquad (4.32)$$

Since the equations are linear, all the fields will have the same dependence on space ($\sim e^{-2x/\ell_s}$) and time of the driving force F_x. We split δn_e and E_x into secular and oscillating components as follows

$$\delta n_e = [\delta n_0 + \text{Re}(\delta \tilde{n} e^{-2i\omega t})]e^{-2x/\ell_s}, \tag{4.33}$$

$$E_x = [E_{x0} + \text{Re}(\tilde{E}_x e^{-2i\omega t})]e^{-2x/\ell_s}. \tag{4.34}$$

In steady conditions and for immobile ions there may be no net drift of the electron fluid, thus for the velocity $u_{x0} = 0$ and

$$u_x(x,t) = \text{Re}(\tilde{u}_x e^{-2i\omega t})e^{-2x/\ell_s}, \tag{4.35}$$

and the secular component of the total force must be vanishing. This brings the balance condition

$$eE_{x0} = F_0, \qquad \delta n_0 = \frac{1}{4\pi e}\left(\frac{-2}{\ell_s}E_{x0}\right) = \frac{F_0}{2\pi e^2 \ell_s}. \tag{4.36}$$

Notice that $\delta n_0 > 0$ since the ponderomotive force piles electron up in the $x > 0$ region. To ensure the overall charge neutrality, a positive surface charge density $\sigma_0 = E_{x0}/4\pi = F_0/4\pi e$ must appear at $x = 0$. This surface density corresponds to an electron depletion layer whose thickness d is much smaller than ℓ_s if $\omega_p \gg \omega$ and $a^2(0) \ll 1$:

$$d = \frac{\sigma_0}{en_0} = \frac{F_0}{4\pi e^2 n_0} = \frac{m_e c^2}{2\ell_s}\frac{a^2(0)}{4\pi e^2 n_0} = \frac{c^2}{2\omega_p^2 \ell_s}a^2(0) \simeq \frac{c}{2\omega_p}a^2(0) \ll \ell_s. \tag{4.37}$$

Thus, taking the depletion layer as a surface charge is consistent with the ordering we assumed.

If the polarization is not circular, we must calculate the oscillating components driven by the "2ω" force. By direct substitution of (4.33–4.35) into (4.32) and some simple algebra we obtain

$$\delta \tilde{n} = n_0 \frac{2F_0/m_e \ell_s}{\omega_p^2 - 4\omega^2}\left(1 + \frac{1-\epsilon^2}{1+\epsilon^2}\right). \tag{4.38}$$

The resonant denominator is due to excitation of plasma oscillations at $\omega_p = 2\omega$ by the driving force. The total electron density is thus given by the sum of secular and oscillating terms

$$\delta n_e = n_0 \frac{2F_0}{m_e \ell_s \omega_p^2}e^{-2x/\ell_s}\left(1 + \frac{1-\epsilon^2}{1+\epsilon^2}\frac{\cos 2\omega t}{1 - 4\omega^2/\omega_p^2}\right). \tag{4.39}$$

Let us consider the case of *linear* polarization, $\epsilon = 1$. In such a case, due to the resonant denominator the amplitude of the oscillating term is larger than that of the secular term. Thus $\delta n_e < 0$ for some time interval, in which there appears a deficiency of electrons in the $x > 0$ region (despite the average piling up caused by the ponderomotive force). This means that in such time interval some electrons have escaped on the vacuum side crossing the $x = 0$ surface. The motion of electrons in the vacuum region cannot be described via the present perturbative approach (since the equations cannot be linearized for small density perturbations). However, as inferred above their dynamics is expected to be similar to that investigated by the capacitor model of Sect. 4.2.2, so that bunches of fast electrons are generated at 2ω rate. Instead, for *circular* polarization ($\epsilon = 1$) the oscillating term in the $-e\mathbf{v} \times \mathbf{B}$ force vanishes and thus we expect fast electron generation to be *suppressed* as there is no force driving electrons across the boundary.[4]

4.2.4 Numerical Simulations and Experiments

The electrostatic modeling of the previous sections is able at most to provide a qualitative picture of fast electron generation in overdense plasmas. Brunel (1988) already performed an investigation of his mechanism using 2D PIC simulations to address electromagnetic effects, the most evident being that electrons acquire also a velocity component parallel to the surface, because of momentum conservation. Later several authors studied absorption and fast electron generation using 1D simulations in the boosted frame (Sect. 2.3.3). A discussion and a comparison of these 1D studies can be found in Gibbon et al. (1999). As an example of such simulations Fig. 4.5 shows three snapshots of the electron density n_e and the longitudinal phase space distribution $f_e(x, p_x)$ for linear polarization (LP) at $\theta = 0°$ and $30°$ and circular polarization (CP) at $\theta = 0°$, respectively. It is apparent that for LP the most energetic ion bunches are generated at a rate of 2ω and ω for normal and oblique incidence, respectively, while for CP fast electron generation is suppressed, being the peak value of p_x nearly two orders of magnitude smaller than for LP.

Simulations performed over a wide range of parameters show that, as a general trend, the absorption degree in fast electrons is quite sensitive to n_0, θ and to the density scalelength $L = n_c/|\partial_x n_e|_{n_e=n_c}$. This is not unexpected because of the already non-trivial dependence on θ and ω_p/ω of Fresnel formulas, the presence of plasma resonances and the importance of initial conditions for the nonlinear electron dynamics. Single particle studies of the latter in the EM fields near the plasma surface have been also performed (see e.g. Bauer et al. 1995; Sentoku et al. 2002) showing in some cases the onset of stochastic behavior and chaos. The electron trajectories are thus strongly dependent on initial conditions, which might account qualitatively for the important effect of a finite gradient at the surface.

Additional complexity appears due to ion motion, that leads to a modification of the density profile already on a scale of tens of femtoseconds, and to multi-dimensional effects. Since the laser pulse has a finite width, the pushing by the

---

[4] A deeper inspection of Eq. (4.39) shows that in the case of elliptical polarization, it is easily found that $\delta n_e < 0$ may occur only if the ellipticity parameter $\epsilon < (\omega_p^2/2\omega^2 - 1)^{-1/2} \simeq \sqrt{2}\omega/\omega_p$. This suggests that there is an "ellipticity threshold" for the onset of fast electron generation by the $\mathbf{v} \times \mathbf{B}$ force.

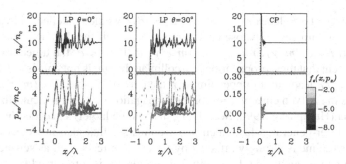

Fig. 4.5 Snapshots of n_e and $f_e(x, p_x)$ (contours in $\log_{10}$ scale) from three 1D PIC simulations for Linear Polarization and incidence angle $\theta = 0°$ and $30°$ and Circular Polarization at $\theta = 0°$. Parameters common to all cases are $n_0 = 10n_c$, $a_0 = 3$ and the time is $t = 7\lambda/c$ after the interaction begins

ponderomotive force of the plasma surface leads to a parabolic deformation of the latter, so that the effective incidence angle changes locally and in time causing a time-dependent absorption (Wilks et al. 1992; Ruhl et al. 1999). The deformation also acts as a geometrical "funnel" favoring the collimation of the fast electrons, an effect that may also be induced by pre-shaping the targets. Recent experiments using targets of such shape have evidenced a strong increase of both the energy and number of fast electrons (Gaillard et al. 2011), allowing for a new record in the maximum energy of ions accelerated in the electron sheath at the rear side (Sect. 5.2).

On the experimental side, there is plenty of evidence for fast electron generation, but precise measurements of the electron distribution function inside the target are obviously difficult and based on indirect methods, such as measuring the line and Bremsstrahlung radiation emitted in collisions with background ions. In addition, the distribution inside the target is also determined by the transport dynamics that may be extremely complex. A crucial point is that the current carried by fast electrons is huge. The discussion on generation mechanism at the interaction surface leads us to infer that at strongly relativistic intensities a typical order of magnitude of the fast electron current density is $J_f \sim en_c c$, typically corresponding to several TA/cm^2. Such a huge current must be neutralized by an opposite "return" current $\mathbf{J}_r$ provided by the background electrons, otherwise the space-charge electric field would grow immediately up to values able to stop the electrons. In fact, if J_f was flowing freely for $t > 0$ from the target surface, a surface charge density $\sigma \sim J_f t$ would there accumulate, generating an uniform electrostatic field $E \sim 4\pi\sigma = 4\pi en_c ct$. At the current front position $x \simeq ct$ the work done by E on an electron would be $eEx \sim 4\pi e^2 n_c c^2 t^2 = m_e c^2 (\omega t)^2$, implying that relativistic electrons may be stopped in a time $\sim \omega^{-1}$, shorter than a laser cycle.

The current neutralization must be local ($\mathbf{J}_f + \mathbf{J}_r \simeq 0$) otherwise the associated magnetic field would inhibit the current propagation as well by deflecting the electron trajectories. For a total current I over a circle of radius a (corresponding to the laser sport radius), the maximum magnetic field at the beam edge would be

$B_{\max} = 2I/ac$. Fast electrons of velocity $u_f = \beta_f c$ would be deflected by the magnetic field on a trajectory whose curvature radius at half the distance from the axis $r_L = u_f/(eB_{\max}/2m_e\gamma_f c) = \beta_f/(eI/m_e c^3\gamma_f a)$. If $r_L < a$, the trajectory of electrons is so bent that their propagation as a collimated beam is not consistent anymore. Posing $a > r_L$ leads to an upper limit for the current, $I < I_A \equiv (m_e c^3/e)\beta_f\gamma_f$. This yields the Alfven current limit for the propagation of a charge neutralized beam in a plasma (Humphries 1990, Sect. 12.7), that can be largely exceeded by the total current of laser-generated fast electrons. We thus see that also fast electron transport in this regime is strongly affected by collective effects such as self-generation of electric and magnetic fields. Additional complexity is introduced by collisional effects on the return currents, refluxing of fast electrons in finite targets, and instabilities. Nevertheless, at least a fraction of fast relativistic electrons is able to propagate coherently through the target conserving the periodic, "pulsed" structure imprinted by the generation mechanism. This has been confirmed by measurements of the optical radiation emitted from the rear surface,[5] that contains components at ω and 2ω depending on the polarization and incidence angle of the driving laser (Popescu et al. 2005).

References

Barnes, W.L., Dereux, A., Ebbesen, T.W.: Nature **424**, 824 (2003)
Bauer, D., Mulser, P., Steeb, W.H.: Phys. Rev. Lett. **75**, 4622 (1995)
Bigongiari, A., Raynaud, M., Riconda, C., Héron, A., Macchi, A.: Phys. Plasmas **18**, 102701 (2011)
Blumenfeld, I., et al.: Nature **445**, 741 (2007)
Brunel, F.: Phys. Rev. Lett. **59**, 52 (1987)
Brunel, F.: Phys. Fluids **31**, 2714 (1988)
Bulanov, S.V., Pegoraro, F., Pukhov, A.M., Sakharov, A.S.: Phys. Rev. Lett. **78**, 4205 (1997)
Bulanov, S.V., Naumova, N., Pegoraro, F., Sakai, J.: Phys. Rev. E **58**, R5257 (1998)
Bulanov, S.V., et al.: In: Shafranov, V.D. (ed.) Reviews of Plasma Physics, vol. 22, p. 227. Kluwer Academic/Plenum Publishers, New York (2001)
Caldwell, A., Lotov, K., Pukhov, A., Simon, F.: Nature Phys. **5**, 363 (2009)
Catto, P.J., More, R.M.: Phys. Fluids **20**, 704 (1977)
Esarey, E., Sprangle, P., Krall, J., Ting, A.: IEEE Trans. Plasma Sci. **24**, 252 (1996)
Esarey, E., Schroeder, C.B., Leemans, W.P.: Rev. Mod. Phys. **81**, 1229 (2009)
Faure, J., et al.: Nature **431**, 541 (2004)
Faure, J., et al.: Nature **444**, 737 (2006)
Gaillard, S.A., et al.: Phys. Plasmas **18**, 056710 (2011)
Geddes, C.G.R., et al.: Nature **431**, 538 (2004)
Gibbon, P.: Short Pulse Laser Interaction with Matter. Imperial College Press, London (2005)
Gibbon, P., et al.: Phys. Plasmas **6**, 947 (1999)
Gonsalves, A.J., et al.: Nature Phys. **7**, 862 (2011)
Grimes, M.K., Rundquist, A.R., Lee, Y.S., Downer, M.C.: Phys. Rev. Lett. **82**, 4010 (1999)
Hafz, N.A.M., et al.: Nature Photon. **2**, 571 (2008)
Hosokai, T., et al.: Appl. Phys. Lett. **96**, 121501 (2010)

---

[5] This emission is due to *transition radiation*, that occurs whenever a charged particles crosses a sharp interface between two media of different refractive index (Jackson 1998, Sect. 13.7).

Humphries, S.: Charged Particle Beams. Wiley, New York (1990)

Jackson, J.D.: Classical Electrodynamics, 3rd edn. Wiley, New York (1998)

Kruer, W.L., Estabrook, K.: Phys. Fluids **28**, 430 (1985)

Landau, L.D., Lifshitz, E.M., Pitaevskii, L.P.: Electrodynamics of Continuous Media, 2nd edn. Elsevier, Butterworth-Heinemann, Oxford (1984)

Leemans, W.P., et al.: Nature Phys. **2**, 696 (2006)

Mangles, S.P.D., et al.: Nature **431**, 535 (2004)

Matlis, N.H., et al.: Nature Phys. **2**, 749 (2006)

Max, C.E.: In: Balian, R., Adam, J. (eds.) Laser-Plasma Interaction, pp. 304–411. North-Holland, Amsterdam (1982, Proc. Les Houches Summer School)

Modena, A., et al.: Nature **377**, 606 (1995)

Mora, P., Antonsen, T.M.: Phys. Rev. E **53**, R2068 (1996)

Mulser, P., Bauer, D.: High Power Laser-Matter Interaction. Springer, Berlin (2010)

Mulser, P., Ruhl, H., Steinmetz, J.: Laser Part. Beams **19**, 23 (2001)

Ozbay, E.: Science **311**, 189 (2006)

Popescu, H., et al.: Phys. Plasmas **12**, 063106 (2005)

Price, D.F., et al.: Phys. Rev. Lett. **75**, 252 (1995)

Pukhov, A., Meyer-ter-Vehn, J.: Appl. Phys. B: Laser Opt. **74**, 355 (2002)

Rozmus, W., Tikhonchuk, V.T., Cauble, R.: Phys. Plasmas **3**(1), 360 (1996)

Ruhl, H., Macchi, A., Mulser, P., Cornolti, F., Hain, S.: Phys. Rev. Lett. **82**, 2095 (1999)

Sentoku, Y., et al.: Appl. Phys. B: Laser Opt. **74**, 207 (2002)

Tajima, T., Dawson, J.M.: Phys. Rev. Lett. **43**, 267 (1979)

Weibel, E.S.: Phys. Fluids **10**, 741 (1967)

Wilks, S.C., Kruer, W.L., Tabak, M., Langdon, A.B.: Phys. Rev. Lett. **69**, 1383 (1992)

Chapter 5
Ion Acceleration

Abstract In this chapter we discuss basic reference mechanisms for ion acceleration in laser-plasma interactions. The first two mechanisms, namely sheath acceleration and plasma expansion, are oriented to the modelization of experiments with solid targets in the so-called *target normal sheath acceleration* (TNSA) framework. The other mechanisms, namely *shock acceleration*, *coulomb explosions* and *radiation pressure acceleration* dominate over TNSA in particular conditions and may allow to develop advanced schemes of ion acceleration.

5.1 Ion Acceleration Scenario

For all the phenomena described in the previous chapters, the dynamics of ions have been ignored because of their larger mass with respect to electrons, so that their contribution can be ignored on fast time scales such as $\sim \omega^{-1}$ or $\sim \omega_p^{-1}$ with ω and ω_p the laser and the (electron) plasma frequency, respectively. Ions typically respond on a time scale $\sim \omega_{pi}^{-1} \sim (m_p/m_e)^{1/2} \omega_p^{-1}$ to slowly varying electric fields generated by large charge separations. The latter are generated by laser-plasma interactions either when large number of high energy "fast" electrons escape from the plasma or by the direct action of the ponderomotive force, as shown in example in the description of self-focusing (Sect. 3.3) and overdense penetration (Sect. 3.4.1). In this way, an indirect transfer of the laser energy to the ions occurs.

The interest of ion acceleration in the superintense regime has been greatly boosted since the year 2000 when three experiments reported on the observation of collimated proton beams with multi-MeV energies from the rear (non-irradiated) side of solid targets (Clark et al. 2000; Maksimchuk et al. 2000; Snavely et al. 2000). The interpretation of these early observations, as well as of a great number of later experiments has stimulated a wide effort in the modeling of the acceleration process. Most of the experiments may be described in the framework of the Target Normal Sheath Acceleration (TNSA) model, which may be briefly described as follows. The fast

A. Macchi, *A Superintense Laser-Plasma Interaction Theory Primer*,
SpringerBriefs in Physics, DOI: 10.1007/978-94-007-6125-4_5,
© The Author(s) 2013

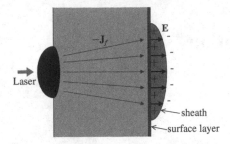

Fig. 5.1 Schematic of the TNSA mechanism. The flow of fast electrons generated at the front side crosses the target bulk and reaches the rear side, generating an electron sheath in vacuum. The electric field is almost perpendicular to the rear surface and accelerates ions, in particular protons from impurity layers on the surface

electrons produced in a high-intensity interaction with the front surface of a solid target propagate through the bulk and eventually reach the rear side, where they produce a negatively charged sheath (see Fig. 5.1). The electrostatic field generated in the sheath is almost normal to the surface and accelerates ions. The protons which are contained in impurity layers, ordinarily present on the surface of metal targets, are in a favorable condition for ion acceleration because they are initially located at the maximum of the sheath field and, thanks to their higher charge-to-mass ratio, they move faster than heavier ions and may screen the electric field. The rapid acceleration from the initially cold surface and from the highly oriented electrostatic field leads to a high collimation and very low emittance of the proton beam, which is very attractive for applications. With respect to traditional accelerators, peculiar features of the laser-accelerated protons also include the very high number of particles for pulse and the very short duration of the latter.

Intense beams of multi-MeV protons and heavier ions have indeed a great potential for many applications where localized energy deposition in matter is required, since the deposited energy has a sharp maximum (the Bragg peak) at the end of their path, differently from electrons and photons. Such behavior is related to the well-known scaling of the Coulomb cross section as $\sim \mathcal{E}_i^{-2}$, with $\mathcal{E}_i$ the ion energy, so that the more energy is deposited the more efficient the deposition becomes. The perspectives for the use of laser-driven ion beams in high energy density science and medicine (e.g. for hadrontherapy) greatly stimulated the research effort after the first experiments. The need to reach both the high energies ($> 100\,\text{MeV}$) and the beam quality required by most applications has further stimulated the search for alternative schemes of ion acceleration, beyond the TNSA model. A detailed account of experimental research, proposed schemes and foreseen applications may be found in recent reviews (Daido et al. 2012; Macchi et al. 2012).

The interpretation of the early observations already stimulated a wide effort in the modeling of the acceleration process. The TNSA picture encompasses different models based either on a static or a dynamic description of the sheath produced by fast electrons. These two regimes are described in the following Sects. 5.2 and

5.4, respectively. Both models are based on an electrostatic, non-relativistic approximation for the ion dynamics which is adequate for present-day regimes where the energy per nucleon is ~ 100 MeV at most. The basis of both models are the fluid equations for ions with electrons assumed in a Boltzmann equilibrium, as described in Sect. 2.2.2. The description is decoupled from the laser-interaction dynamics; it is assumed that the latter has created an electron population of given density and temperature, which are taken as initial conditions.

The same electrostatic fluid equations are used in Sect. 5.5 to describe shock acceleration, a possibly emerging topic. The emphasis will be on the dynamics of nonlinear shock waves in a collisionless plasma, which accelerate ions by reflection at the shock front.

The regime of Coulomb explosions (Sect. 5.6) occurs in situations when electron are completely removed from a finite size plasma and ions are accelerated by their own space-charge field. The description of the acceleration process is thus greatly simplified by the presence of a single species.

Differently from the above mechanisms, radiation pressure acceleration (RPA) is driven directly by the radiation pressure exerted by superintense laser pulses on overdense plasmas. For RPA (Sect. 5.7) we turn back to a relativistic description because of both its apparent potential for the production of relativistic (> 1 GeV energy) ions and the connection with the physics of "moving mirrors" whose applications will be further discussed in Chap. 6.

5.2 Sheath Acceleration

Multi-MeV protons were observed in the experiments of Clark et al. (2000), Maksimchuk et al. (2000), Snavely et al. (2000) both using plastic (CH) and metal targets, in which hydrogen-containing impurities are present on the surface. The interpretation of these observations was the subject of intense debate and controversy, but eventually it appeared that the dominant contribution to proton emission came from the rear side of the target. In the TNSA model originally proposed in Wilks et al. (2001), fast electrons (Sect. 4.2) accelerated at the front side and crossing the bulk of the target enter the vacuum region at the rear side, where a charged sheath is formed. The electric field in the sheath back-hold electrons and accelerates ions outwards. In the simplest picture, the fast electron sheath may be described in the assumption of Boltzmann equilibrium, and assumed to remain static during the acceleration of the fastest ions at least. Protons from impurities are thus in a favorable condition because they are initially located at the edge of the target, i.e. at the maximum of the sheath field, and will move faster than all other ions thanks to their higher charge-to-mass (Z/A) ratio.

Assuming the density profile at the rear side to be infinitely sharp (i.e. ion density $n_i = (n_0/Z)\Theta(-x)$), the simple modeling of the electron sheath is analogous to that in Sect. 4.2.1 but now using parameters for *fast* electrons, i.e. their temperature T_f and density n_f. Their typical values may be estimated from direct measurements of the electron spectra and the absorption coefficient A or, in the case such measurements are unavailable, using scaling laws and absorption models, see e.g. Eqs. (4.10) and (4.29). From the knowledge of T_f and A and assuming a quasi-steady state one may write a balance condition between the incoming flux of absorbed laser power AI and the outcoming flux of energy carried by fast electrons, $n_f T_f u_f$ where u_f is the electron velocity ($\simeq c$ for highly relativistic electrons). Posing $AI \simeq n_f T_f u_f$ yields an estimate for n_f in the production region (and thus $n_f \lesssim n_c$ is usually taken as an additional condition for consistency). However, the fast electron density inside the target may change due to electron beam divergence and refluxing. In any case, T_f is mostly relevant in order to estimate the energy at which the ions are accelerated because the potential drop in the sheath will be $\Delta \Phi_s \sim T_f / e$ in order to back-hold the electrons. Hence, a "test" ion of charge Z moving in such potential will gain an energy $\mathcal{E}_i \simeq Ze\Delta\Phi_s \simeq ZT_f$.

We may expect to provide a more precise estimate by solving the equations describing a static sheath.[1] Assuming a step-like ion density as above and a Boltzmann relation (2.72) for electrons, Poisson's Eq. (2.73) becomes

$$d_x^2 \Phi = 4\pi e(n_e - Zn_i) = 4\pi e n_0 \left[e^{e\Phi/T_e} - \Theta(-x) \right]. \quad (5.1)$$

If the plasma is globally neutral, the integral from $-\infty$ to $+\infty$ of (5.1) must be zero, yielding the condition $d_x\Phi(-\infty) = d_x\Phi(+\infty) = 0$. In addition, the electron density should remain equal to the background value deeply inside the plasma, so $n_e(-\infty) = n_0$, and vanish in vacuum far from the surface, so $n_e(+\infty) = 0$. Thus for the electrostatic potential we have $\Phi(-\infty) = 0$ and $\Phi(+\infty) = -\infty$. These relations show a serious drawback that is an unavoidable consequence of the use of the Boltzmann relation (2.72): the potential drop from 0 to ∞ is infinite, and so is also the energy acquired by a test ion moving in such potential.

A possible way out of this difficulty is to assume that the range of electron energies is not infinite but has some upper cut-off $\bar{\mathcal{E}}$, that seems physically reasonable in general. In such a case the binding potential $\bar{\mathcal{E}}$ will be limited with $\Phi(+\infty) = -\bar{\mathcal{E}}$ that also gives an upper limit to the maximum energy gain of test ions. Notice that an energy distribution "truncated" to some upper value will always come out for an isolated, warm plasma of *finite* size in 3D, which acquires a net charge Q because the escape energy for electrons has a finite value. If the plasma fills a sphere of radius R, the escape energy for an electron will be $U_{\text{esc}} = eQ/R = e^2 N_{\text{esc}}/R$ with N_{esc} the number of escaped electrons. Still assuming the electrons to follow a Boltzmann distribution, we may estimate $N_{\text{esc}} = N_0 \exp(-U_{\text{esc}}/T_e)$ with $N_0 = n_0(4\pi R^3/3)$ the total number of electrons. Thus, N_{esc} (and U_{esc}) may be determined from the implicit equation

---

[1] In order to present the equations in a more general form, we use T_e for the electrons temperature; when applying the formulas to fast electrons, $T_e = T_f$ should be posed. It is worth reminding, however, that we do not consider the dynamics of cold background electrons, which slightly affect sheath formation.

$$\log(N_{esc}/N_0) = -e^2 N_{esc}/(RT_e) = -(R/\lambda_D)^2(N_{esc}/N_0)/3. \qquad (5.2)$$

The charging of laser-irradiated targets is actually a transient phenomenon, that has been observed with high spatial and temporal resolution (see e.g. Quinn et al. 2009); in the typical configuration of TNSA experiments, neutralizing currents will flow on the target surface from a region around the sheath whose radius increases at the speed of light. The system is thus in general not in an equilibrium state, and it can be considered neither electrically grounded nor isolated. These issues make the question of boundary conditions for static models, which attempt to make accurate descriptions of the electric sheath, a difficult one. A discussion can be found in Macchi et al. (2012) and references therein.

We now look for solutions of Eq. (5.1) in the vacuum region $x \geq 0$. Switching to dimensionless quantities $\xi = x/\lambda_D$, $\phi = e\Phi/T_e$, $E = eE_x\lambda_D/T_e$ for convenience, (5.1) becomes $d_\xi^2\phi = e^\phi - \Theta(-\xi)$ (notice that e = 2.718... is Euler's number!). Multiplying both sides by $E = -d\phi/d\xi$ we obtain

$$(d_\xi \phi)(\partial_\xi^2 \phi) = d_\xi (d_\xi \phi)^2/2 = d_\xi \left(E^2/2 \right) = \left(e^\phi - \Theta(-\xi) \right) d_\xi \phi. \qquad (5.3)$$

Integrating over ξ from 0 to $+\infty$ and then from $-\infty$ to 0 yields the two relations

$$E^2(0)/2 = -e^{\phi(0)}, \qquad E^2(0)/2 = -e^{\phi(0)} + \phi(0) + 1. \qquad (5.4)$$

For consistency we thus obtain $\phi(0) = -1$ and $E(0) = \sqrt{2/e}$, i.e. the field at the surface is given by

$$E_x(0) = \sqrt{2/e}E_0, \qquad E_0 = \sqrt{4\pi n_0 T_e}. \qquad (5.5)$$

The values of $\phi(0)$ and $E(0)$ can be used as boundary conditions to obtain the solution for $x < 0$ numerically, while an analytical solution can be found for $x > 0$. To do this in the equation $d_\xi^2\phi = e^\phi$ we put $F = e^\phi$ from which we obtain $d_\xi F = \sqrt{2}F^{3/2}$ i.e. $F^{1/2}(\xi) - F^{1/2}(0) = -\xi\sqrt{2}$, and with some algebra we eventually obtain (Crow et al. 1975)

$$\phi = -2\ln\left(\xi/\sqrt{2} + e\right), \qquad E = 2/(\xi + \sqrt{2}e). \qquad (5.6)$$

The electric field profile is shown in Fig. 5.2.

Now, if we assume that the electron energy distribution has a cut-off at an energy uT_e, with $u > 0$, it is still possible to write down an analytical solution in which ϕ does not diverge but becomes constant at some distance $x_r = \xi_r\lambda_D$ from the surface, so that $E = 0$ for $x \geq x_r$. Integrating (5.1) from $\xi = 0$ to ξ_r, we now obtain $E^2(0)/2 = e^{\phi(0)} - e^{-u}$, and $(d_\xi\phi)^2/2 = e^\phi - e^{-u}$, that yields $d_\xi\phi = -\sqrt{2}(e^\phi - e^{-u})^{1/2}$. Still posing $F = e^\phi$ with some algebra[2] we eventually obtain for the potential (Passoni and Lontano 2004)

---

[2] The indefinite integral we need is $\int x^{-1}(x-a)^{-1/2}dx = 2\arctan\left[(x-a)^{-1/2}\right]$ for $a > 0$.

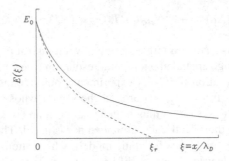

Fig. 5.2 The sheath field in the vacuum region, assuming a standard Boltzmann distribution with temperature T_e [Eq. (5.6), *thick line*] and a "truncated" distribution with maximum electron energy uT_e [Eq. (5.8), *dashed line*], for which $E = 0$ for $x > x_r$. (In the figure $u = 3$ has been assumed)

$$\phi = -u + \ln\left(1 + \tan^2\left((\xi - \xi_r)e^{-u/2}/\sqrt{2}\right)\right), \tag{5.7}$$

and for the electric field

$$E = -\sqrt{2}e^{-u/2}\tan\left((\xi - \xi_r)e^{-u/2}/\sqrt{2}\right), \tag{5.8}$$

where $\xi_r = \sqrt{2}e^{+u/2}\arctan(E(0)e^{+u/2}/\sqrt{2})$ gives the finite range of the sheath: for $\xi \geq \xi_r$, $\phi = -u$ and $E = 0$. The "truncated" field profile is also shown in Fig. 5.2 for $u = 3$.

The description of a static sheath may be improved by, e.g., more accurate considerations on the distributions of fast electrons (see e.g. Passoni et al. 2010 and references therein). Actually the latter is the main source of uncertainty when comparing to experimental observations. Despite such difficulty, static models may reproduce the observed scalings for the cut-off energy of protons with good accuracy and a limited set of parameters, suggesting that the approximation of a static sheath is a fair one at least for the fastest protons.

If the initial density profile is not sharp, the sheath field E_s becomes weaker, as might be qualitatively understood by estimating $eE_s \sim T_f/L_s$ with L_s the scalelength of the sheath, so that $E_s \ll E_0$ (5.5) when $L_s \gg \lambda_D$. This is consistent with the experimental observation that preforming a plasma of finite density gradient at the rear surface destroys proton acceleration (Mackinnon et al. 2001). On the other hand, using very thin (sub-micrometric) targets and ultrashort laser pulses with very high contrast to prevent early plasma formation, fast electrons escape also from the front side while the density profile remains sharp also there, so that a rather symmetrical proton emission is observed on both sides of the target (Ceccotti et al. 2007).

5.3 Plasma Expansion

To go beyond the static approximation of Sect. 5.2, we describe TNSA as the expansion of an electron-ion plasma in vacuum, a classic problem of plasma physics. We keep the Boltzmann equilibrium assumption for electrons, despite its above discussed drawbacks, and thus use Eqs. (2.72) and (5.1) with (2.67) and (2.71) to describe the ion dynamics. We make the additional assumption that the plasma is *locally* neutral, i.e. $\varrho = 0$ and thus $n_e = n_i$. This does not imply that the electrostatic field is absent: indeed, we assume that the spatial scales of charge separation layers are of the order of the local Debye length $\lambda_D = (T_e/4\pi e^2 n_i)^{1/2}$, so that the plasma appears to be neutral on a coarser scale. Notice also that for the 1D case in the region of "quasi-neutrality" there may exist a *uniform* field E_x, which is a solution of the homogeneous 1D Poisson's equation and whose sources are outside the quasi-neutral region, i.e located where the $n_e = n_i$ assumptions may break down. This is indeed the case for the nonlinear solution we describe below, which corresponds to the well-known isothermal rarefaction wave (Zel'dovich and Raizer 1966; Gurevich et al. 1966) that has been used since longtime to describe the expansion of laser-produced plasmas (Gitomer et al. 1986).

By posing $n_i = n_e = n_0 \exp(e\Phi/T_e)$ and differentiating with respect to x we obtain

$$\partial_x n_i = n_0(e/T_e)\left(e^{e\Phi/T_e}\right)\partial_x \Phi = -(e/T_e)n_i E_x, \qquad (5.9)$$

so that by replacing for E_x in (2.71) for ions

$$(\partial_t - u_i\partial_x)u_i = -(ZT_e/m_i)\partial_x n_i/n_i = -c_s^2\partial_x n_i/n_i, \qquad (5.10)$$

where $c_s = (ZT_e/m_i)^{1/2}$ is the speed of sound or, more precisely, the *ion-acoustic* velocity. In fact, linearizing (5.10) and (2.67)

$$\partial_t u_i \simeq -c_s^2\partial_x n_i/n_0 , \qquad \partial_t n_i \simeq -n_0\partial_x u_i, \qquad (5.11)$$

and looking for plane wave solutions $\sim\exp(ikx-i\omega t)$ we readily obtain the dispersion relation $\omega^2 = k^2 c_s^2$, describing non-dispersive ion-acoustic waves with velocity c_s.

Now, Eqs. (5.10) and (2.67) also have a nonlinear *self-similar* solution, for which each variable is a function of $s = x/t$ only, i.e. $n_i = N(s)$ and $u_i = V(s)$. To prove this we first notice that

$$\partial_t (u_i, n_i) = -\left(x/t^2\right)(V', N'), \qquad \partial_x (u_i, n_i) = (1/t)(V', N'), \qquad (5.12)$$

where $(V', N') = d(V, N)/ds$. Substituting in (5.10) and (2.67) for ions yields

$$V' = -(N'/N)(V - s), \qquad N'/N = -(V - s)/c_s^2. \qquad (5.13)$$

Eliminating (N'/N) the equation for V becomes $V' = (V - s)^2/c_s^2$, which has a solution $V = c_s + s$, $V' = 1$, corresponding to a velocity linearly increasing in x

$$u_i = u_i(x, t) = c_s + x/t. \tag{5.14}$$

To describe a simple expansion of the plasma we must have $u_i > 0$, thus (5.14) is valid only for $x > -c_s t$. For $x < -c_s t$, ions are not yet in motion ($u_i = 0$) and thus the density must be unperturbed, $n_i = n_0$. Thus, $x = -c_s t$ defines a rarefaction front moving backwards inside the target at the speed of sound c_s. After substituting for $V - s = c_s$, the equation for N becomes $N'/N = -1/c_s$, with solution $N(s) = N(0) \exp(-s/c_s)$, i.e.

$$n_i(x, t) = n_0 e^{-x/(c_s t)-1}. \tag{5.15}$$

The electric field for $x > -c_s t$ can be obtained from (5.9) as

$$E_x = -(T_e/e)\partial_x n_i/n_i = 2E_0/\omega_{pi}t, \tag{5.16}$$

and is thus uniform as we anticipated. Since $E_x = 0$ in the unperturbed target ($x < -c_s t$), a surface charge density $\rho_1 = (E_x(t)/4\pi)\delta(x + c_s t)$ must exist at the rarefaction front. This positive charge must be balanced by an opposite amount of negative charge located somewhere in the $x > -c_s t$ region, so that the neutral solution must become invalid at some point, as also indicated by the unlimited increase of u_i with x. To determine the front where the neutrality assumption breaks down we compare the local density scalelength, $L \equiv n_i/|\partial_x n_i|$, with the local Debye length λ_D. In fact, inside a plasma in Boltzmann equilibrium space-charge fields are screened at distances larger than λ_D, so that charge separation occurs over distances smaller than λ_D and in any case where the density varies on a scale shorter than λ_D. Since $L = c_s t$, by posing $\lambda_D = \lambda_D(x/t) = c_s t$ we find the condition $x = x_f(t)$ where

$$x_f(t) = c_s t[2 \ln(\omega_{pi}t) - 1], \qquad u_f(t) = d_t x_f = c_s[2 \ln(\omega_{pi}t) + 1]. \tag{5.17}$$

It is reasonable that the density profile does not extend up to $x = +\infty$ since this would imply that ions with infinite velocity exist. A sketch of the density and velocity profiles obtained by truncating the self-similar solution at $x = x_f(t)$ is shown in Fig. 5.3.

Notice that the truncated profile does not satisfy the conservation of particle number, since

$$\int_{-c_s t}^{+\infty} n_i(x, t)dx = -n_0 c_s t\, e^{-x/c_s t-1}\Big|_{-c_s t}^{x_f} = n_0 c_s t - n_0 c_s/(\omega_{pi}^2 t) < n_0 c_s t \,, \tag{5.18}$$

that is the total number of ions (per unit surface) which have been set in motion at the time t. Thus, near the ion front the actual ion density must be higher than the

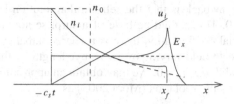

Fig. 5.3 Sketch of the ion density (n_i), velocity (u_i) and electric field (E_x) profiles in the 1D isothermal expansion of a step-boundary plasma. *Dashed* and *continuous lines* show the self-similar and the actual "corrected" profiles, respectively

self-similar value. We also notice that the field at the ion front $E_f = E_f(t)$, which we can obtain from the acceleration as

$$E_f = E_x(x = x_f) = (m_e/e)d_t u_f = (m_e/e)(2c_s\omega_{pi}/\omega_{pi}t) = 2E_0/\omega_{pi}t, \quad (5.19)$$

that is twice the field from the self-similar solution. Thus, we expect the exact field at the front (where charge separation effects can not be neglected) to have a peak, jumping from the self-similar value to approximately twice of that, and then decreasing down to zero in the sheath region beyond the front. These qualitatively inferred features of n_i and E_x are also sketched in Fig. 5.3 and are in agreement with the numerical solution of the hydrodynamical equations (Mora 2003).

The condition $x = x_f(t)$ defines the front of the fastest ions moving at velocity $u_f = u_f(t)$ and thus gives also the high-energy cut-off in the energy spectrum. To determine the latter at a given time t, we notice that $n_i = n_0 \exp(-u_i/c_s)$ with $u_i \leq u_f$ holds, so that in terms of the energy $\mathcal{E}_i = m_i u_i^2/2$ we obtain

$$\frac{dn_i}{d\mathcal{E}_i} = \frac{dn_i}{du_i}\frac{du_i}{d\mathcal{E}_i} = \frac{n_0/T_e}{\sqrt{2\mathcal{E}_i/T_e}} \exp\left(-\sqrt{\frac{2\mathcal{E}_i}{T_e}}\right), \qquad \mathcal{E}_i \leq \mathcal{E}_{\max}(t) = \frac{m_i}{2}u_f^2. \quad (5.20)$$

The spectrum is plotted in Fig. 5.4.

Fig. 5.4 The instantaneous energy spectrum (5.20) $dn_i/d\mathcal{E}$ (normalized to n_0/T_e), for a cut-off energy $\mathcal{E}_f = 2.8T_e$

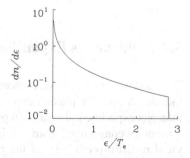

A first apparent drawback is that the self-similar solution becomes singular at the initial time $t = 0$. This should be little of a surprise now because we already knew that in the initial configuration there is a charge unbalance at the surface of the plasma with a sheath field varying over a Debye length, so that the quasi-neutral solution cannot be valid near $t = 0$. We may obtain an approximate solution for the field at the ion front by interpolation (Mora 2003) as

$$E_f(t) \simeq 2E_0/(2\mathrm{e} + \omega_{pi}^2 t^2)^{1/2}, \tag{5.21}$$

so that at $t = 0$ the field equals the peak field in the static sheath given by (5.5). The corresponding ion front velocity u_f is then obtained by integrating the equation of motion $d_t u_f = (Ze/m_i)E_f$ with $u_f(0) = 0$, obtaining

$$u_f(t) \simeq 2c_s \ln \left[\frac{\omega_{pi} t}{\sqrt{2\mathrm{e}}} + \left(\frac{\omega_{pi}^2 t^2}{2\mathrm{e}} + 1 \right)^{1/2} \right], \tag{5.22}$$

that has the same asymptotic limit as the second of Eqs. (5.17).

The second drawback is that the maximum velocity of ions, and hence the cut-off energy, diverges logarithmically with time. This is an unavoidable consequence of the isothermal assumption: the system has an infinite energy reservoir in the electron fluid and thus it is able to accelerate ions indefinitely. There is no easier way to remove this unphysical behavior from the model but to give a constraint of finite energy (per unit surface), for instance by assuming the plasma to be of finite thickness, and consistently to have an electron temperature $T_e = T_e(t)$ decreasing in time due to cooling (see e.g. Betti et al. 2005), all at a cost of more involved analytical calculations. Nevertheless, the simplicity of Eq. (5.22) has proven to be attractive and difficult to abandon, thus it has been suggested to insert a phenomenological "maximum acceleration time" t_m at which the acceleration should stop. In the interpretation of old experiments with long nanosecond pulses t_m has been roughly estimated as the laser duration (Gitomer et al. 1986). More recently, in experiments with superintense sub-picosecond pulses t_m has been used as a parameter to fit experimental data (Fuchs et al. 2006) with the goals of estimating sheath plasma parameters from the observed ion energy cut-off and of inferring scaling laws for $\mathcal{E}_{\max}$.

5.4 Multispecies Expansion and Monoenergetic Acceleration

The modeling of the previous Sections assumed a single ion species. Actually, both the sheath and the plasma expansion models have been used in the literature to interpret experimental results on proton acceleration from targets where hydrogen is a minority component. Thus, we infer that the static sheath model (Sect. 5.2) might yield realistic predictions if the protons can be considered as test ions, i.e. their number is insufficient to modify the sheath field and their dynamics is faster than the typical time for sheath evolution. For what concerns the expansion model (Sect. 5.3), the number of protons should be indeed high enough to screen the total charge of the fast electrons, so that acceleration of heavier species is inhibited.

A considerable research effort has been pursued in order to obtain narrow energy spectra, as required by most foreseen applications. A strategy has been to use special targets in which the protons (or a different species to be accelerated) are concentrated in a narrow layer at the rear surface, so that the initial conditions are the same for all protons and the same final energy would be then expected. Actually, in metallic targets hydrogen impurities are already in a very narrow layer at the rear surface, thus one may wonder why broad, exponential-like spectra are typically observed in normal conditions. First, the sheath field has a sharp peak at the density discontinuity, thus its variation in longitudinal direction may be relevant even for a very thin surface layer. Second, the sheath field is inhomogeneous in the transverse direction, because of the laser intensity distribution in the focal spot, and in fact a correlation between the spectral width and the angular direction in which protons are detected is observed. Relatively narrow proton spectra have been obtained using small microdot coatings on the rear surface, so that the spatial variation over the proton emitting area is reduced (Schwoerer et al. 2006). Narrow band spectra have been also produced for Carbon ions in targets where an ultrathin graphite layer replaced hydrogen-containing impurities (Hegelich et al. 2006).

Actually, the formation of peaks in the energy spectrum is strongly dependent on the relative densities of the species present in a composite target, and typically occurs during the expansion of a multispecies plasma. To provide a simple and, to some extent, heuristic description of this effect let us still consider the case of a semi-infinite, isothermal plasma but with two ion species, namely protons and heavy ions with $Z/A \ll 1$. If the initial density of the protons $n_{p0} \ll n_{h0}$, the density of heavy ions, then we may expect the plasma expansion to be dominated by the heavier species, so that the latter is described by the self-similar solutions (5.14–5.16) where $n_0 \to n_{h0}$ and $c_s \to c_h = (ZT_e/Am_p)^{1/2}$. For protons, we assume that they are rarefied enough to give a negligible perturbation to the field and thus can be treated as test particles which move in the uniform field created by the heavy ions:

$$m_p(\partial_t - u_i\partial_x)u_i = 2eE_0/\omega_{ph}t = T_e/c_h t. \tag{5.23}$$

The equation of motion for protons is thus identical to (5.10) but with the electric field $E_x = T_e/c_h t > T_e/c_p t$, the field in the case of single species expansion $(c_p = (T_e/m_p)^{1/2})$. The protons thus gain a relatively higher energy than in the single species case. The equations of motion have self-similar solutions analogous to (5.14, 5.15), i.e.

$$u_p = c'_p + x/t, \quad n_p = n_{p0}e^{-x/(c'_p t)-1}, \quad c'_p = c_p^2/c_h, \tag{5.24}$$

where for the moment we do not care about boundary conditions at the heavy ions rarefaction front $x = -c_h t$, where the self-similar field drops to zero. Indeed, we notice that the proton density decreases on a distance $c_p't > c_h t$, so that at some point $n_p > n_h$, violating the starting assumption of small proton density. The condition $n_p = n_h$ marks the boundary of a transition region beyond which the expansion becomes dominated by the protons. Assuming that (5.24) holds, $n_p = n_h$ at $x \simeq x_{tr}(t) \equiv (c_p' c_h / (c_p' - c_h)) t \log(n_{h0}/n_{p0})$ where protons have thus the velocity $u_p' \simeq c_p^2/c_h + c_h \log(n_{h0}/n_{p0})$ if $c_p \gg c_h$, that does not depend on time. Beyond this front, the electric field drops down, so that there is a modest further energy gain for protons as they cross the transition region; this corresponds to a *plateau* in the velocity profile and to a peak at $\sim m_p(u_p')^2/2$ in the energy spectrum. Far away enough from the transition region we expect heavy ions to be absent and the proton velocity to be thus described by the single species solution $u_p = c_p + x/t$, with $c_p < c_p'$. A rough estimate for the end point of the transition region may thus be obtained by matching the two solutions, i.e. $u_p' := c_p + x/t$.

A more extended analysis, supported by numerical simulations, may be found in Tikhonchuk et al. (2005) that contains also references to earlier work. To obtain an approximate solution which satisfies the boundary condition $u_p = 0$ at the rarefaction front $x = -c_h t$, we assume $x/t \ll c_p$. In this way, the equations for the self-similar variables are simplified as $V V' = -c_p^2/c_h$ and $V N' = -N V'$. The second equation implies that the proton flux NV is a constant. The solutions of these approximated equations may be written as

$$u_p \simeq c_p\sqrt{2}(1 + x/c_h t)^{1/2}, \qquad n_p \simeq (n_{p0} c_h / c_p \sqrt{2})(1 + x/c_h t)^{-1/2}. \quad (5.25)$$

Since the flux is constant and the density varies smoothly, protons will tend to accumulate in the transition region, forming a spectral peak in correspondence of the plateau velocity. The position at which $n_p = n_h$ is now roughly estimated as $x_{tr} \simeq c_h t \log\left(\sqrt{2}(c_p/c_h)(n_{h0}/n_{p0})\right)$ from which one estimates the plateau velocity $u_p' = u_p(x_{tr})$ (notice that $x_{tr}/t \ll c_p$ is required for consistency with the starting assumptions). The energy at the peak is

$$\mathcal{E}_p' = \frac{m_p}{2}(v_p')^2 \simeq T_e \log\left(\sqrt{\frac{2A}{Z}}\frac{n_{h0}}{n_{p0}}\right), \quad (5.26)$$

where we used $c_p/c_h = A/Z$ and $m_p c_p^2 = T_e$, Despite the roughness of the calculations presented, the important message is that playing with the relative proton concentration n_{p0}/n_{h0} and the charge-to-mass ratio Z/A of the heavy species, a monoenergetic peak whose energy increases with decreasing proton density might be obtained. In addition, the mechanism of species separation with formation of spectral peaks is quite general and thus may be always at play in composite targets.

Recently, narrow ion spectra have been observed experimentally also in particular interaction conditions, i.e. using very thin targets for which relativistic transparency

(Sect. 3.4) occurs during the interaction, leading to strong heating of electrons. The observations are described in Hegelich et al. (2011) and Jung et al. (2011) where references to related theoretical models may also be found.

5.5 Collisionless Shock Acceleration

In hydrodynamics, shock waves (or, briefly, shocks) appear as discontinuous solutions, with fluid variables such as the density making a jump at a discontinuity surface (the shock front) moving at a velocity V_s (the shock velocity). In a broad sense, a shock implies a change of the macroscopic state of the medium across a propagating transition layer. In a collisionless plasma, shock-like solutions can be found for the same system of equations used in Sect. 5.3. Such *collisionless electrostatic shocks* (CES) propagate with Mach numbers $M = V_s/c_s > 1$ and are characterized by the potential Φ making a positive jump at the shock front. Thus, plasma ions initially ahead of the shock front may be reflected by the moving potential barrier depending on their initial energy and on the barrier height. If the ions are initially at rest, elastic reflection from the moving barrier leads to a final velocity of $2V_s$, thus the shock front is preceded by a flow of ions having twice its velocity.

The idea of collisionless shock acceleration (CSA) in a laser plasma is based on the guess that the presence of a population of fast electrons may result in high values of c_s, hence in high energy of the reflected ions $m_i(2V_s)^2/2 = 2m_iV_s^2 = 2ZM^2T_h$. The shock may be driven at the laser-plasma interaction surface by the piston action of radiation pressure (Silva et al. 2004) (see also Sect. 5.7). The piston velocity u_p may be estimated by a flow momentum balance $m_i(n_iu_p)u_p \sim I/c$, yielding a typical velocity

$$u_p = \left(\frac{I}{m_in_ic}\right)^{1/2} = \left(\frac{Z}{A}\frac{m_e}{m_p}\frac{n_e}{n_c}\right)^{1/2}a_0c. \tag{5.27}$$

In a hot plasma and if $u_p > c_s$ a shock propagating into the plasma at $V_s \simeq u_p$ might be generated. Shocks may be also driven by fast electron heating, which creates a strong gradient in the energy density. As long as V_s is constant, the spectrum of reflected ions should be monoenergetic. In a recent experiment using CO_2 laser pulses and overdense hydrogen gas targets, narrow spectra were observed and attributed to CSA (Haberberger et al. 2012). This scheme might be promising for applications also because the combination of gas target and gas laser allows for high repetition rate operation.

We review the basic theory of CES in a plasma, following the classic book of Tidman and Krall (1971, Chap. 6). As it will turn out, the existence of reflected ions may be necessary to the formation of CES (rather than being a consequence of CES propagation), while in the case of no reflection solitary wave solutions, i.e. *solitons* are found.

We use again the ES system of Eqs. (2.67, 2.71–2.73) with $\mathbf{B} = 0$ and $\mathsf{P}_i = 0$ and we search for shock wave-like solutions characterized by a constant front velocity V in the laboratory frame, with the front position separating "upstream" ($x > Vt$) and "downstream" ($x < Vt$) regions. We assume the waves to be stationary in the frame moving at velocity $V > 0$ with respect to the laboratory. In such frame $\partial_t n_i = 0$ and $\partial_t u_i = 0$, thus (2.67, 2.71) give

$$\partial_x(n_i u_i) = 0, \qquad u_i \partial_x u_i = -(Ze/m_i)\partial_x \Phi. \tag{5.28}$$

In the moving frame, ions in the upstream region and far from the transition region, i.e. at $x \to +\infty$, move at velocity $-V$ towards the front. Thus, integration of the above equations across the wave front gives

$$n_i u_i = -(n_0/Z)V, \qquad m_i u_i^2/2 + Ze\Phi = m_i V^2/2, \tag{5.29}$$

where $\Phi = 0$ ahead of transition layer has been assumed. Eliminating u_i we obtain n_i as a function of Φ,

$$n_i = \frac{n_0}{Z} \frac{V}{(V^2 - 2Ze\Phi/m_i)^{1/2}}, \tag{5.30}$$

that is real valued only if $m_i V^2/2 > Ze\Phi$. This condition is equivalent to state that the background ions have enough kinetic energy to overcome the potential barrier and are *not* reflected by the wave front.

To keep the notation compact from now on we switch to dimensionless units posing $\xi = x/\lambda_D$, $\phi = e\Phi/T_e$ and $M = V/c_s$. By substituting for n_i Poisson's equation $\partial_x^2 \Phi = 4\pi e(n_e - Zn_i)$ we obtain

$$\frac{d^2\phi}{d\xi^2} = \left[e^\phi - \frac{M}{(M^2 - 2\phi)^{1/2}} \right], \tag{5.31}$$

which, after multiplication by $d_\xi \phi$ and integration, yields $(d_\xi \phi)^2/2 + U_1(\phi) = 0$, where

$$U_1(\phi) = -\left[e^\phi + M(M^2 - 2\phi)^{1/2} \right] + 1 + M^2, \tag{5.32}$$

and we choose the integration constant in order that $U_1(0) = 0$. Equation (5.31) is thus identical to Newton's equation for a particle moving in a potential U_1, as it is apparent when replacing $\xi \to t$ and $\Phi \to q$, so that the trajectory $q(t)$ is equivalent to $\Phi(\xi)$. We will use this analogy to discuss the solutions to Eq. (5.31). This approach is known as the Sagdeev *pseudopotential* method.

First we remind that $\Phi < m_i V^2/2Ze$, i.e. $\phi < M^2/2 \equiv \phi_{cr}$ and we notice that $(dU_1/d\phi)(0) = 0$. Thus, since ϕ is a continuous function, we infer that $U_1(0^+) < 0$ and $U_1(\phi_{cr}) > 0$ are sufficient conditions to state that there is a zero of $U_1(\phi)$ at some value $\phi_m < \phi_{cr}$ and that $U_1(\Phi)$ has a minimum in the $[0, \phi_m]$ interval. In this way, U_1 has the form shown in Fig. 5.5a. According to Eq. (5.32), the pseudoparticle

Fig. 5.5 a Pseudopotential
functions U_1 (5.32) and U_2
(5.35). **b** Profiles of $\phi(x)$ for
a soliton solution (*dashed*),
corresponding to the motion
of a pseudoparticle starting
from $\phi = 0$ in U_1, and a
shock solution (*thick*) where
the pseudoparticle "jumps"
into U_2 at $\phi = \phi_m$

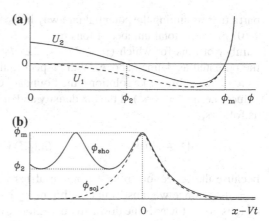

has zero energy, thus it bounces into the "pseudopotential" well between $\phi = 0$ and
$\phi = \phi_m$. However, since $(dU_1/d\phi)(0) = 0$, the particle leaves and reaches the $\phi = 0$
position "in an infinite time", i.e. it performs a *single oscillation*. The corresponding
solution for $\phi(x)$ has the form of a solitary wave or *soliton* and is shown in Fig. 5.5b.

The conditions $U_1(0^+) < 0$ and $U_1(\phi_{cr}) > 0$ determine the range of parameters
for which the soliton exists. To check the first condition, we expand $U_1(\phi)$ in Taylor
series around $\phi = 0$ up to second order $U_1(\Phi) \simeq -(\phi^2/2)\left(1 - M^{-2}\right) :< 0$ that
yields the condition $M^2 > 1$, i.e. the soliton is supersonic. For what concerns the
second condition, we substitute $\phi = \phi_{cr} = M^2/2$ in $U_1(\phi)$ and obtain $e^{M^2/2} - M^2 - 1 < 0$, that can be solved numerically to yield $M < 1.5852\ldots \simeq 1.6$.

Soliton solutions are symmetric and the fluid state is restored when passing
through the soliton, in contrast to what is expected for a shock wave. To obtain
a shock wave solution, a way to "break the symmetry" must be looked for. The center position ξ_m of the soliton corresponds to the peak value $\phi_m = \phi(\xi_m)$ for which
$U(\phi_m) = 0$ (from now on we pose $\xi_m = 0$). The idea is to modify the boundary conditions in the upstream region, $x > 0$, so that the pseudopotential function
becomes a different function when crossing the $x = 0$ position. In the pseudoparticle
analogy, the potential is instantaneously changed at the time when the particle has
zero velocity and invert its direction. It is sufficient, roughly speaking, to lift up the
"new" pseudopotential U_2 near $\phi = 0$ so that $U_2(0) > 0$, in order that now the
pseudoparticle is trapped into a potential well and an *oscillating* solution is obtained.

Allowing for a fraction of the ions to be reflected at the shock front brings the
desired symmetry breaking for the pseudopotential. Let the background ions have a
velocity distribution function $f_i = f_i(v)$, which may be, e.g., a drifting Maxwellian
in the shock frame, $f_i \propto \exp[-(v + V)/u_{ti}]$. The function

$$F(\Phi) = \int_{(v+V)^2 < 2Ze\Phi/m_i} f_i(v)dv \tag{5.33}$$

gives, as a function of Φ, the number of ions which are reflected back at some value of
Φ, because their kinetic energy is not sufficient to overcome the potential barrier. In

particular, assuming the potential far away in the upstream region $\Phi(x = +\infty) = 0$, $F(0)$ gives the total number of ions reflected by the potential at some point, i.e. the number of ions for which $(v + V)^2 < 2Ze\Phi_m/m_i$, being Φ_m the peak value of the potential as defined above. With respect to the above derivation which brought us to obtain soliton-like solutions, the boundary conditions defining the background values (i.e. at $x = +\infty$) of the ion density and of the ion stream current are modified as follows,

$$n_0 \to n_0[1 + F(0)], \qquad n_i u_i = -(n_0/Z)V \to (n_0/Z)[1 + F(0)]V, \qquad (5.34)$$

because the reflection from the potential barrier increases the number of ions at $x = +\infty$ (since we must count the bouncing back ions) and decreases the number of ions going towards the downstream region. Correspondingly, the average density decreases to $n_0[1 - F(0)]$ in the downstream region ($x < 0$). Poisson's equation (5.31) assumes a different form in the $x > 0$ and $x < 0$ regions:

$$\frac{d^2\phi}{d\xi^2} = \left[(1 + S(\xi)F(0))e^\phi - \frac{M(1 - F(0))}{(M^2 - 2\phi)^{1/2}} - 2\Theta(\xi)F(\phi) \right] \equiv -\frac{dU_2}{d\phi}, \quad (5.35)$$

where $S(\xi) = \pm 1$ and $\Theta(\xi) = 1$ or 0 for $\xi > 0$ and $\xi < 0$, respectively. The additional term $2\Theta(\xi)F(\phi)$ in the upstream region is the local contribution of reflected ions. It can be seen that such symmetry breaking yields a pseudopotential U_2 of the desired form, yielding a shock-like solution. In the downstream region, ϕ now oscillates between the peak value ϕ_m and the minimum value ϕ_2 corresponding to the second zero of the pseudopotential, $U(\phi_2) = 0$ with $0 < \phi_d < \phi_m$, see Fig. 5.5.

Simulations show that in intense laser-plasma interactions either solitons or shocks may be generated depending on the initial velocity distribution of background ions, that also determines the number of reflected ions. If the latter is too high the shock may decelerate and cause the spectrum to broaden (Macchi et al. 2012). Therefore, for CSA the efficiency of a single shot is probably not compatible with keeping a monoenergetic spectrum. For applications, this limitation might be compensated by high repetition rate operation.

5.6 Coulomb Explosions

If the target has a limited size, typically smaller than a laser wavelength, a laser pulse may be intense enough to blow out all the electrons in the target. Subsequently, ions are accelerated by their own space-charge field, that is actually the maximum electrostatic field that can be generated for a target of given size and density. This is the so-called *Coulomb Explosion* (CE) regime that may be most easily realized by irradiating small clusters which are formed in low-temperature gas jets.

To describe the CE of a spherical, cold ($T_i = 0$) cloud of N ions with mass m_i and charge $q = Ze$, let us indicate with $r_i = r_i(t)$ the position of a given layer

Fig. 5.6 a Coulomb explosion of a *spherical* plasma
cloud. For cold ions, the
charge contained inside the
$r = r_i(t)$ sphere is constant.
b The final spectrum of ions
as a function of the maximum
energy $\mathcal{E}_m$, Eq. (5.37)

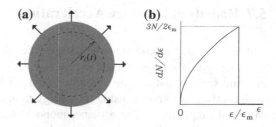

(shell), with $0 < r_i(0) < R$ being R the initial radius of the spherical cloud (see
Fig. 5.6). Because of Gauss's theorem the radial electrostatic field $E(r)$ is a growing
function of r, hence outer layers have higher acceleration than inner layers and ions
initially at some radius $r_1(0)$ never overturn those initially at $r_2(0) > r_1(0)$. Thus,
if $r_i = r_i(t)$ is the position of a spherical layer at the time t, the charge inside the
$r = r_i(t)$ sphere is constant and thus equal to the initial value $Q_i = qN(r_i(0)/R)^3$.
Hence using Gauss's theorem the electric field $E[r_i(t)] = Q_i/r_i^2(t)$ (as if the charge
Q_i was concentrated in the centre of the sphere) so that the equation of motion for
the layer is

$$m_i \frac{d^2 r_i(t)}{dt^2} = \frac{Ze Q_i}{r_i^2(t)} = \frac{(Ze)^2 N}{r_i^2(t)} \left(\frac{r_i(0)}{R}\right)^3, \qquad (5.36)$$

and thus the ions in the layer with initial radius $r_i(0)$ eventually gain the energy
$\mathcal{E} = (Ze)^2 N r_i^2(0)/R^3$ with maximum value $\mathcal{E}_m = (Ze)^2 N/R$. The energy spectrum
may be obtained as

$$\frac{dN}{d\mathcal{E}} = \frac{dN/dr_i(0)}{d\mathcal{E}/dr_i(0)} = \frac{3N}{2} \frac{\mathcal{E}^{1/2}}{\mathcal{E}_m^{3/2}}, \qquad \mathcal{E} \le \mathcal{E}_m, \qquad (5.37)$$

being $dN = 4\pi n_0 r_i^2(0) dr_i(0)$ with $n_0 = N/(4\pi R^3/3)$. The spectrum is shown in
Fig. 5.6b. An experimental demonstration of the high energy obtained in the CE of
small ($N = 10^2 - 10^6$, $R \lesssim 10^{-2}\,\mu m$) Deuterium clusters even with a moderate
energy pulse was provided by the observed high rate of nuclear fusion reactions
(Ditmire et al. 1999; Zweiback et al. 2000).

Using the same approach as above it is easy to describe CE in both planar (slab) and
cylindrical (wire) geometry where, however, the final energy of ions diverges because
in both cases the spatial extension and the energy of the system are effectively infinite.
Nevertheless, the formulas may be useful to describe at least the initial stages of ion
acceleration in a cavitation channel generated in underdense plasmas (see Sect. 3.3,
Macchi et al. (2009) and references therein) or in an ultrathin foil where the laser
pulse blows all electrons away. In the latter case, the boundary conditions are such
that the electric field is null at the front side and peaks at the rear side, resulting in a
"directed" CE (Bulanov et al. 2008; Grech et al. 2009).

5.7 Radiation Pressure Acceleration

EM waves carry momentum, which may be delivered to either an absorbing or a reflecting medium. Radiation pressure (RP) is the flow of delivered momentum per unit surface and can be computed (at least in principle) when the EM fields at the surface are known, using the energy-momentum conservation theorem from Maxwell's equations.[3] For a plane, monochromatic EM wave of intensity I and frequency ω normally incident on the plane surface of a medium ("target") at rest, the RP P_{rad} is given by

$$P_{\text{rad}} = (1 + R - T)I/c = (2R + A)I/c , \tag{5.38}$$

where R and T are the reflection and transmission coefficients of the target as defined in the derivation of Fresnel formulas. Energy conservation implies $R + T = 1 - A$ where A is the absorption coefficient, i.e. the fraction of the EM wave intensity that is converted into internal energy. Limiting cases include a perfect mirror ($R = 1$, $T = A = 0$, $P = 2I/c$), a perfectly absorbing medium ($A = 1$, $R = T = 0$, $P = I/c$) and a "thin" transmitting, non-absorbing target such that $A = 0, T = 1-R$, $P = 2RI/c$.

In the present case it is instructive to show explicitly that the RP on a step-boundary overdense plasma ($n_e = n_0\Theta(x)$) is due to the ponderomotive force (PF). In fact, considering normal incidence and the non-relativistic case for simplicity, the PF is the cycle average of (4.31) and thus the total pressure is the integral of the volume force over the depth of the plasma:

$$P_{\text{tot}} = \int_0^{+\infty} n_0 F_0 e^{-2x/\ell_s} dx = F_0 n_0 \ell_s /2 = (m_e c^2 n_0)(\omega/\omega_p)^2 a_0^2 = 2I/c, \tag{5.39}$$

where we used $I = m_e c^3 n_c a_0^2$ and $(\omega/\omega_p)^2 = n_c/n_0$.

As also seen in Sect. (4.2.3), the PF pushes and piles up electrons in the skin layer creating a static field $E_{0x} = f_p/e$ that acts on the ions, so that effectively the RP is exerted on the whole target. In a thick target, ions are thus accelerated at the front surface causing steepening of the density profile and, for a finite pulse in multi-dimensional geometry, *Hole boring* (HB) through the plasma. If the target is thin enough and has a low mass, it may be accelerated as a whole up to high velocity: this is suggestively named the *Light sail* (LS) regime. In the following we describe these two operational modes of Radiation pressure acceleration (RPA)[4] by assuming for

---

[3] The electromagnetic theory of radiation pressure is due to Maxwell (1873). It is however interesting that the Italian physicist Adolfo Bartoli also obtained independently Maxwell's result in 1875 from thermodynamic considerations: see Bartoli (1884).

[4] The reader should be warned that in the literature the regimes of RPA have been named in various ways. The HB regime has been often described as "electrostatic shock acceleration", while we consider this definition appropriate for the acceleration mechanism described in Sect. 5.5. The term "laser piston" has been used both for RPA of thin foils (Esirkepov et al. 2004) and of thick targets (Schlegel et al. 2009). "Sweeping acceleration" has been also used to describe HB acceleration in

simplicity that the laser pulse is plane and circularly polarized, so that effects of laser-driven oscillations and electron heating may be neglected; in a more general case such effects overlap to those associated to RP. Before that, we evaluate the radiation pressure on a "moving mirror" as it will be used for both HB and LS modeling.

5.7.1 The Moving Mirror

We evaluate the RP on a reflecting surface moving at a velocity $V = \beta c$ in the laboratory frame S. We assume that the surface is a "moving mirror" whose reflectivity R is known as a function of the frequency *in the frame S' where it is at rest* (we take $A = 0$, i.e. neglect absorption for simplicity). The RP is the same in S and S' because the force component parallel to the Lorentz boost velocity and the area of the surface where the force is exerted are both invariant. Thus we may write $P_{rad} = P'_{rad} = (2I'/c)R(\omega')$ with I' and ω' the intensity and frequency in S', respectively. The electric field of the incident wave is $E'_0 = \gamma(E_0 - \beta B_0/c) = \gamma(1-\beta)E_0$ since $E_0 = B_0$, and thus $I' = cE'^2_0/4\pi = I(1-\beta)/(1+\beta)$. The frequency of both the incident and reflected waves in S' is $\omega' = \gamma(1-\beta)\omega = ((1-\beta)/(1+\beta))^{1/2}$ because of the Doppler effect. Thus we find

$$P_{rad} = \frac{2I}{c}R(\omega')\frac{1-\beta}{1+\beta}, \qquad \omega' = \omega\left(\frac{1-\beta}{1+\beta}\right)^{1/2}. \qquad (5.40)$$

It may be instructive to obtain this result also in a different way. For simplicity we assume a perfect mirror, i.e. $R = 1$. From a quantum point of view the RP originates from the reflection of a number N (per unit surface) of photons with energy-momentum $(\hbar\omega, \hat{x}\hbar\omega/c)$ contained in a short bunch of duration τ, corresponding to an intensity $I = N\hbar\omega/\tau$. The number N is conserved in the reflection and it is an invariant, while the frequency of the reflected photons ω_r in S' gains another Doppler factor with respect to ω':

$$\omega_r = \omega'\left(\frac{1-\beta}{1+\beta}\right)^{1/2} = \omega\frac{1-\beta}{1+\beta}. \qquad (5.41)$$

The time during which reflection occurs in the laboratory is $\tau_r = \tau/(1-\beta)$, where the denominator originates from the motion of the surface during the reflection. The resulting pressure is thus $P_{rad} = |\Delta\mathbf{p}|/\Delta t = (N\hbar/c)(\omega+\omega_r)/\tau_r$ which brings (5.40) for $R = 1$. This argument should make clear that the relativistic factor in (5.41) allows to account self-consistently for the EM energy loss to the acceleration of matter.

---

the density profile at the front of a target (Sentoku et al. 2003). A particular condition, or feature, of LS has been described as "phase stable acceleration" (Yan et al. 2008). Here we focus on the basic "thin" (LS) and "thick" (HB) regimes without addressing details of RPA-based schemes.

5.7.2 Hole Boring Regime

In the hole boring (HB) regime the recession velocity of the plasma surface u_{hb} may be simply estimated by balancing the EM and mass momentum flows in a planar geometry as previously done to obtain Eq. (5.27). Here we generalize the balance equation for the relativistic case in which u_{hb} may be close to c (Robinson et al. 2009; Schlegel et al. 2009). Considering the laser pulse to act as a piston or snowplow on the plasma, in S' we observe a flow of incoming ions with velocity $-u_{hb}$ which reach the surface and, in steady conditions, are bounced back with velocity u_{hb}. In S' the ion density is $n_i \gamma_{hb}$ with $\gamma_{hb} = (1 - u_{hb}^2/c^2)^{-1/2}$ because of the relativistic contraction. The EM momentum flow, i.e. the RP, must then balance a momentum flow difference equal to $n_i \gamma_{hb} (2m_i \gamma_{hb} u_{hb}) u_{hb}$. Using (5.40) the balance equation thus gives

$$\frac{2I}{c} \frac{1 - u_{hb}/c}{1 + u_{hb}/c} = n_i \gamma_{hb} (2m_i \gamma_{hb} u_{hb}) u_{hb}. \tag{5.42}$$

Solving for u_{hb} this yields

$$\frac{u_{hb}}{c} = \frac{\Pi^{1/2}}{1 + \Pi^{1/2}}, \qquad \Pi = \frac{I}{m_i n_i c^3} = \frac{Z}{A} \frac{n_c}{n_e} \frac{m_e}{m_p} a_0^2. \tag{5.43}$$

The fastest ions are those bouncing back from the surface in the moving frame, resulting in a maximum energy per nucleon in the lab frame

$$\mathcal{E}_{\max} = 2m_p c^2 \frac{\Pi}{1 + 2\Pi^{1/2}}. \tag{5.44}$$

In the non-relativistic regime where $\Pi \ll 1$ and $u_{hb} \ll c$, we obtain $u_{hb}/c \simeq \Pi^{1/2}$ and $\mathcal{E}_{\max} \simeq 2m_p c^2 \Pi$.

The above model assumes steady conditions and does not resolve the dynamics of ion acceleration in the charge separation layer. To provide a simple picture of the latter we use simplified profiles of density and fields as shown in the cartoon of Fig. 5.7 (Macchi et al. 2005). The first frame, Fig. 5.7a describes the initial equilibrium for immobile ions: the electrons have piled up under the action of the PF that is balanced by the space-charge field E_x. The model parameters x_s, x_d, E_0 and n_{p0} (defined in Fig. 5.7a) are related to each other by $E_0 = 4\pi e n_0$ (from Poisson's equation), $n_{p0}(x_s - x_d) = n_0 x_s$ (from charge conservation) and $e E_0 n_{p0} = 2I/c$ (from the balance between the total electrostatic pressure $-e \int n_e E_x dx$ and the RP). A fourth relation to close the system may be obtained by noticing that $\ell_s = x_s - x_d$ is the penetration distance of the ponderomotive force into the target, e.g. $\ell_s = c/2\omega_p \gg x_d$ for the linearized model of Sect. 4.2.3. Now we follow the motion of the ions initially in the $x_s < x < x_d$ region (in the non-relativistic approximation for simplicity). Using Lagrange coordinates (Sect. 4.2.2), we notice that, as far as the ion trajectories do not intersect, the electric field on each layer depends only on the initial position (due

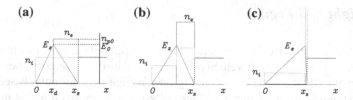

Fig. 5.7 Model profiles of the ion density n_i, electron density n_e and electrostatic field E_x at three stages of HB acceleration: **a** the initial equilibrium, **b** during acceleration, **c** at the formation of a singularity in n_i

to Gauss's theorem in 1D), so that the equation of motion is

$$m_i \frac{d^2 x_i(t)}{dt^2} = ZeE_{x0}, \qquad E_{x0} = E_0 \left(1 - \frac{x_i(0) - x_d}{\ell_s}\right). \qquad (5.45)$$

The simple solution $x_i(t) = x_i(0) + (ZeE_{x0}/2m_i)t^2$ shows that the density and field profiles in the compression region are self-similar as in Fig. 5.7b, and that all these ions get to the point $x = x_s$ *at the same time* $t_b = (2\ell_s m_i/ZeE_0)^{1/2}$, *independently on the initial position* $x_i(0)$. Hence a singularity appears in the ion density n_i at $x = x_s$ and $t = t_b$, as sketched in Fig. 5.7c. Ions initially in the depletion layer $0 < x < x_d$ are accelerated by their space-charge field as in a directed CE (Sect. 5.6) and do not reach other ions before $t = t_b$.

According to the model, at $t = t_b$ the velocity spectrum of ions with $x_d < x_i(0) < x_s$ is a flat-top distribution extending from zero to the cut-off value $2u_{hb}$ given by (5.43) for $\Pi \ll 1$ (notice that the velocity is independent of ℓ_s). Actually, simulations show that the "final" energy spectrum observed in the simulations is determined by the highly transient stage of equilibrium collapse and "wave-breaking" which follows the formation of the singularity. The fastest ions form a narrow bunch of velocity $\simeq 2u_{hb}$ that penetrates into the overdense plasma detaching from the surface layer, which moves at an average velocity $\simeq u_{hb}$ (Macchi et al. 2005).[5] As long as the laser pulse is not over, after the wave breaking and the ion bunch acceleration a quasi-equilibrium condition is established again and the process repeats itself. HB acceleration is thus of pulsed nature, but the steady model gives correct formulas for the averaged motion.

Although HB may be considered as ubiquitous as the RP action in intense laser interaction with overdense plasmas, several factors such as plasma heating or the modest ion energy expected for solid densities according to (5.44) make its experimental observation not easy. An indication of HB acceleration was given in an experiment employing CO_2 laser pulses and a hydrogen gas jet with density of a few times n_c (Palmer et al. 2011). The observed ion energies were fairly consistent with a scaling $\mathcal{E}_{max} \sim I/n_0$ in agreement with (5.44) for $\Pi \ll 1$.

---

[5] A similar dynamics is observed for radial ponderomotive acceleration in an underdense plasma (Macchi et al. 2009).

5.7.3 Light Sail Regime

The simplest model for LS assumes the target as a rigid plane mirror of finite mass M and area A, moving with velocity $V = dX/dt$ in the laboratory frame S and boosted by the RP of a plane wave (see Fig. 5.8). Intriguingly, formulas based on this model first appeared in a paper by Marx (1966) who proposed a rocket driven by an Earth-based mirror as a promising concept for interstellar travel, because of the foreseen high efficiency. Combining (5.40) with the relativistic equation of motion for $P = M\gamma V = M\gamma\beta c$, we obtain

$$\frac{d}{dt}(\gamma\beta) = \frac{2I(t - X/c)}{\sigma c^2} R(\omega')\frac{1 - \beta}{1 + \beta}, \qquad \frac{dX}{dt} = \beta c . \qquad (5.46)$$

where $\sigma = M/A$ is the surface density. Notice that the intensity of the plane wave propagating in x direction at the location of the sail, $x = X$, is a function of $t - X/c$. It would be wrong to put the intensity dependence $I = I(t)$ in (5.46) since the sail is moving, thus any change in the intensity at the laboratory time t will influence the sail motion only at the "retarded" time $w = t - X(t)/c$. Curiously, such a mistake was made in Marx's paper, but did not affect the final results: the story is discussed in Simmons and McInnes (1993).

Equation (5.46) may be integrated for simple expressions of $R(\omega')$ as that given in Sect. (3.4.2) (Macchi et al. 2010) but here we restrict to the $R = 1$ case for simplicity. First, we notice that $d_t(\gamma\beta) = \gamma^3 d_t\beta$ and then we switch to w as the integration variable, with $dw = dt - d(x(t)/c) = (1 - \beta)dt$. Eq. (5.46) can thus be rewritten as

$$\frac{d}{dw}\left(\frac{1 + \beta}{1 - \beta}\right)^{1/2} = \frac{\gamma}{1 - \beta}\frac{d\beta}{dw} = \frac{2I(w)}{\sigma c^2}, \qquad (5.47)$$

and integration between 0 and w yields

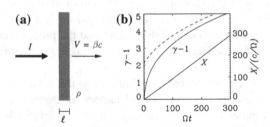

Fig. 5.8 a a simple "accelerating mirror" concept for LS-RPA. **b** The normalized kinetic energy $\gamma(t) - 1$ [Eq. (5.53)] and the position $X(t)$ (normalized to c/Ω) for the LS model with constant intensity. The parameter Ω is defined in Eq. (5.52). The *dashed line* shows the asymptotic $\sim t^{1/3}$ scaling

$$\left(\frac{1+\beta(w)}{1-\beta(w)}\right)^{1/2} - 1 = \frac{2}{\sigma c^2}\int_0^w I(w')dw' \equiv \frac{2F(w)}{\sigma c^2} \equiv \mathcal{F}(w). \tag{5.48}$$

By $F(w)$ we indicate the fluence of the EM wave, i.e. the total *energy* per unit surface that has reached the sail at the time w. Solving for $\beta(w)$ we obtain

$$\beta(w) = \frac{[1+\mathcal{F}(w)]^2 - 1}{[1+\mathcal{F}(w)]^2 + 1}. \tag{5.49}$$

Equation (5.49) is very useful since it yields the final velocity of the sail as a function of the total fluence $\mathcal{F}_\infty$, i.e. by taking $w \to \infty$ to ensure that all of the EM wave pulse has reached the sail, without the need to calculate the explicit trajectory. The corresponding energy per nucleon is given by

$$\mathcal{E}_{max} = m_p c^2 [\gamma(\infty) - 1] = m_p c^2 \frac{\mathcal{F}_\infty^2}{2[\mathcal{F}_\infty + 1]}. \tag{5.50}$$

It is also of interest to calculate the instantaneous efficiency η, i.e. the ratio between the variation of the sail's energy at the (retarded) time w divided by the electromagnetic power delivered to the sail at the same (retarded) instant, i.e. by $I(w)$:

$$\eta \equiv \frac{d_w(\sigma c^2 \gamma)}{I(w)} = \frac{\sigma c^2}{I(w)}\left(\beta\gamma^3\frac{d\beta}{dw}\right) = \frac{\sigma c^2}{I(w)}\beta\gamma^3\frac{1-\beta}{\gamma}\frac{2I(w)}{\sigma c^2} = \frac{2\beta}{1+\beta}, \tag{5.51}$$

where (5.47) has been used. This result also follows from an argument based on the conservation of the number of photons in the reflection from the sail (see Sect. 5.7.1). Consider N photons, carrying a total energy $N\hbar\omega$, being reflected during a short time interval such that β does not vary significantly. The total energy of the reflected photons is $N\hbar\omega_r$, with ω_r given by (5.41), thus the energy transferred to the mirror is $N\hbar(\omega - \omega_r) = [2\beta/(1+\beta)]N\hbar\omega$ so that the efficiency is $\eta = 2\beta/(1+\beta)$. Notice that $\eta \to 1$ for $\beta \to 1$, i.e. for $\mathcal{F}_\infty \to \infty$.

Analytical solutions for the mirror trajectory can be obtained assuming a constant intensity $I(t) = I\Theta(t)$, for which $\mathcal{F}(w) = 2Iw/\sigma c^2 \equiv \Omega(t - X/c)$, where

$$\Omega = \frac{2I}{\sigma c^2} = \frac{Z}{A}\frac{m_e}{m_p}\frac{a_0^2}{\zeta}\omega, \tag{5.52}$$

with ζ defined in (3.53). Using (5.49) and $dX/dw = c\beta/(1-\beta)$, we integrate over w obtaining $X = (c/2)((1+\Omega w)^2 - 1)$. This relation becomes a cubic equation for $X(t) = c(t-w)$ that can be solved, and from $X(t)$ also $\beta(t) = c^{-1}dX/dt$ and $\gamma(t)$ are obtained (see Fig. 5.8). We omit details and give the final expression for $\gamma(t)$:

$$\gamma(t) = \sinh(u) + \frac{1}{4\sinh(u)}, \quad u \equiv \frac{1}{3}\text{asinh}(3\Omega t + 2). \tag{5.53}$$

Asymptotically, $\gamma(t) \simeq (3\Omega t)^{1/3}$. Thus, the time t_{acc} needed to reach a given energy $\mathcal{E}_{\text{max}} = m_p c^2 (\gamma(t_{\text{acc}}) - 1)$ in the laboratory may be estimated by equating (5.53) and (5.50). Noticing that $\mathcal{F}_\infty = 2\pi \Omega \tau_p$ with τ_p the pulse duration, in the ultrarelativistic limit one obtains $t_{\text{acc}} \simeq (\pi^3/3)\Omega^2 \tau_p^3 = (\pi/12)\mathcal{F}_\infty^2 \tau_p$.

It is simple, and possibly useful, to write down solutions valid to first order in β, for which the LS equation of motion becomes $d_t \beta \simeq \Omega(1 - 2\beta) + \mathcal{O}(\beta^2)$, valid for $\beta < 1/2$. The equation is easily solved to yield $\beta \simeq (1 - \exp(-2\Omega t))/2$ and $X \simeq ct/2 + (c/4\Omega)[\exp(-2\Omega t) - 1]$. These expressions confirm that the energy gain is faster in the non-relativistic limit.

At this point it is natural to ask whether the model of a rigid mirror is adequate to describe the interaction of an intense laser pulse with a thin foil. The analysis of Sect. 5.7.2 shows that a charge separation layer is generated by the ponderomotive force, and that ions are effectively divided in two groups, with the ions in the skin layer being bunched by the electric field. In a thick target these latter ions leave the skin layer and are not accelerated anymore since the laser field is screened by the background plasma. A target can be considered as "thin" when the evanescence point ($x = x_s$ in Fig. 5.7) coincides with the rear edge of the foil, so that the laser may follow the ion bunch and repeat the acceleration stage. An analysis of such "repeated acceleration" mechanism shows that it eventually converges to the "accelerated mirror" description (Grech et al. 2011). Notice that the number of ions coherently accelerated as a "sail" may be much less than the total number of ions in the target and thus the accelerated mass (per unit surface) $\sigma_{\text{acc}} \ll \sigma$, the total mass of the foil. With reference to Fig. 5.7, if the foil thickness $\ell \simeq x_s$, then $\sigma_{\text{acc}} \simeq \sigma(1 - x_d/x_s) \ll \sigma$ occurs when $x_s - x_d \ll x_s \simeq \ell$. Then it may appear surprising that the motion of the sail is well described by Eq. (5.46) where the *total* mass is used. The reason is that the ion layer is actually pushed by the electrostatic pressure *on ions* P_i, that for a non-neutral layer (because electrons are in excess in the $x_s > x > x_d$ region of Fig. 5.7) is *less* than the pressure *on electrons* $P_e = P_{\text{rad}} = 2I/c$. For the profiles in Fig. 5.7 we obtain

$$P_i = \int_{x_d}^{x_s} Ze n_0 E_x dx = \int_{x_d}^{x_s} e E_x n_e \left(1 - \frac{x_d}{x_s}\right) dx = \left(1 - \frac{x_d}{x_s}\right) P_e . \quad (5.54)$$

To obtain the equation of motion for the ion layer one thus should replace σ_{acc} for σ and $P_i = (1 - x_d/x_s)P_{\text{rad}}$ for P_{rad} in (5.46), so that the factor $(1 - x_d/x_s)$ on both sides cancels out. The motion of the "reduced mass" sail is thus still described by (5.46), although the number of ions effectively accelerated coherently (and producing a monoenergetic distribution) may be much less than the total number (Macchi et al. 2009). Further details on the dynamics beneath the LS model may be found in Macchi et al. (2010), Grech et al. (2011), Eliasson et al. (2009).

The onset of foil transparency destroys LS acceleration, thus using the model of Sect. 3.4.2 the energy gain is maximized for $a_0 \simeq \zeta$. This optimal condition is well accessible to present experiments since thin foil manufacturing technology enables to obtain values of ζ in the 1–10 range corresponding to σ of the order of 10^{-6} g cm^{-2}, so that $\mathcal{F}_\infty \simeq 1$ is obtained with fluences of $\simeq 5 \times 10^7$ J cm^{-2}. These figures along with the favorable scaling (5.50) and the foreseen high monoenergeticity and efficiency make LS an attractive scheme for ion acceleration, as was highlighted in several theoretical works (see e.g. Esirkepov et al. 2004; Zhang et al. 2007; Klimo et al. 2008; Robinson et al. 2008). Experimental demonstration of LS, however, has proven to be difficult so far mostly because of excessive heating and deformation of thin foil targets with realistic, tightly focused laser pulses. Nevertheless, recently the fast scaling $\mathcal{E}_{max} \sim \mathcal{F}_{\infty}^2$ for $\mathcal{F}_{\infty} \ll 1$ has been demonstrated (Kar et al. 2012) for both protons and light ($Z/A = 1/2$) ions up to $\mathcal{E}_{max} \simeq 10$ MeV per nucleon with relatively narrow, albeit still significant energy spread. In order to obtain relativistic energies, achieving the necessary acceleration length may appear challenging because of the slow energy gain ($\sim t^{1/3}$). However, theoretical studies suggests that the transverse expansion of the foil might reduce the effective surface density of the foil, leading to a faster gain ($\sim t^{3/5}$) (Bulanov et al. 2010) and to a peak energy that is *higher* for 3D geometry and a pulse of finite width than for the 1D, plane wave case (Tamburini et al. 2012). These features might allow a potentially "unlimited" acceleration at the cost of a smaller number of accelerated ions.

References

Bartoli, A.: Il Nuovo Cimento **15**, 193 (1884)
Betti, S., Ceccherini, F., Cornolti, F., Pegoraro, F.: Plasma Phys. Controlled Fusion **47**, 521 (2005)
Bulanov, S.S., et al.: Phys. Rev. E **78**, 026412 (2008)
Bulanov, S.V., et al.: Phys. Rev. Lett. **104**, 135003 (2010)
Ceccotti, T., et al.: Phys. Rev. Lett. **99**, 185002 (2007)
Clark, E.L., et al.: Phys. Rev. Lett. **84**, 670 (2000)
Crow, J.E., Auer, P.L., Allen, J.E.: J. Plasma Phys. **14**, 65 (1975)
Daido, H., Nishiuchi, M., Pirozhkov, A.S.: Rep. Prog. Phys. **75**, 056401 (2012)
Ditmire, T., et al.: Nature **398**, 489 (1999)
Eliasson, B., Liu, C.S., Shao, X., Sagdeev, R.Z., Shukla, P.K.: New J. Phys. **11**, 073006 (2009)
Esirkepov, T., Borghesi, M., Bulanov, S.V., Mourou, G., Tajima, T.: Phys. Rev. Lett. **92**, 175003 (2004)
Fuchs, J., et al.: Nat. Phys. **2**, 48 (2006)
Gitomer, S.J., et al.: Phys. Fluids **29**(8), 2679 (1986)
Grech, M., Skupin, S., Nuter, R., Gremillet, L., Lefebvre, E.: New J. Phys. **11**, 093035 (2009)
Grech, M., Skupin, S., Diaw, A., Schlegel, T., Tikhonchuk, V.T.: New J. Phys. **13**, 123003 (2011)
Gurevich, A.V., Pariiskaya, L.V., Pitaevskii, L.P.: Sov. Phys. JETP **22**, 449 (1966)
Haberberger, D., et al.: Nat. Phys. **8**, 95 (2012)
Hegelich, B.M., et al.: Nature **439**, 441 (2006)
Hegelich, B., et al.: Nucl. Fusion **51**, 083011 (2011)
Jung, D., et al.: Phys. Rev. Lett. **107**, 115002 (2011)
Kar, S., et al.: Phys. Rev. Lett. **109**, 185006 (2012)
Klimo, O., Psikal, J., Limpouch, J., Tikhonchuk, V.T.: Phys. Rev. ST Accel. Beams **11**, 031301 (2008)
Macchi, A., Borghesi, M., Passoni, M.: Rev. Mod. Phys. (2013, To be published)
Macchi, A., Cattani, F., Liseykina, T.V., Cornolti, F.: Phys. Rev. Lett. **94**, 165003 (2005)
Macchi, A., Ceccherini, F., Cornolti, F., Kar, S., Borghesi, M.: Plasma Phys. Control Fusion **51**, 024005 (2009)

Macchi, A., Veghini, S., Pegoraro, F.: Phys. Rev. Lett. **103**, 085003 (2009)
Macchi, A., Veghini, S., Liseykina, T.V., Pegoraro, F.: New J. Phys. **12**, 045013 (2010)
Macchi, A., Nindrayog, A.S., Pegoraro, F.: Phys. Rev. E **85**, 046402 (2012)
Mackinnon, A.J., et al.: Phys. Rev. Lett. **86**, 1769 (2001)
Maksimchuk, A., Gu, S., Flippo, K., Umstadter, D., Bychenkov, V.Y.: Phys. Rev. Lett. **84**, 4108 (2000)
Marx, G.: Nature **211**, 22 (1966)
Maxwell, J.C.: A Treatise on electricity and magnetism, vol. 2. Macmillan, London (1873)
Mora, P.: Phys. Rev. Lett. **90**, 185002 (2003)
Palmer, C.A.J., et al.: Phys. Rev. Lett. **106**, 014801 (2011)
Passoni, M., Lontano, M.: Laser Part. Beams **22**, 163 (2004)
Passoni, M., Bertagna, L., Zani, A.: New J. Phys. **12**, 045012 (2010)
Quinn, K., et al.: Phys. Rev. Lett. **102**, 194801 (2009)
Robinson, A.P.L., Zepf, M., Kar, S., Evans, R.G., Bellei, C.: New J. Phys. **10**, 013021 (2008)
Robinson, A.P.L., et al.: Plasma Phys. Control Fusion **51**, 024004 (2009)
Schlegel, T., et al.: Phys. Plasmas **16**, 083103 (2009)
Schwoerer, H., et al.: Nature **439**, 445 (2006)
Sentoku, Y., Cowan, T.E., Kemp, A., Ruhl, H.: Phys. Plasmas **10**, 2009 (2003)
Silva, L.O., et al.: Phys. Rev. Lett. **92**, 015002 (2004)
Simmons, J.F.L., McInnes, C.R.: Am. J. Phys. **61**, 205 (1993)
Snavely, R.A., et al.: Phys. Rev. Lett. **85**, 2945 (2000)
Tamburini, M., Liseykina, T.V., Pegoraro, F., Macchi, A.: Phys. Rev. E **85**, 016407 (2012)
Tidman, D.A., Krall, N.A.: Shock Waves in Collisionless Plasmas. Wiley/Interscience, New York (1971)
Tikhonchuk, V.T., Andreev, A.A., Bochkarev, S.G., Bychenkov, V.Y.: Plasma Phys. Controlled Fusion **47**, B869 (2005)
Wilks, S.C., et al.: Phys. Plasmas **8**, 542 (2001)
Yan, X.Q., et al.: Phys. Rev. Lett. **100**, 135003 (2008)
Zel'dovich, Y.B., Raizer, Y.P.: Physics of Shock Waves and High-Temperature Phenomena. Academic Press, New York (1966)
Zhang, X., Shen, B., Li, X., Jin, Z., Wang, F.: Phys. Plasmas **14**, 073101 (2007)
Zweiback, J., et al.: Phys. Rev. Lett. **84**, 2634 (2000)

Chapter 6
Photon Acceleration and Relativistic Engineering

Abstract In this chapter we describe laser-plasma interaction schemes in which the plasma has the role of an active optical medium that increases the frequency of the incident laser pulse, allowing its compression in time and space. The two main schemes we describe, namely "flying mirrors" in underdense plasmas and high harmonic generation in overdense plasmas, are broadly related to the concept of the relativistic moving mirror.

6.1 Towards Higher Frequencies and Higher Fields

Efficient conversion of laser light into higher frequencies (i.e. "photon acceleration") up to the X-ray range and beyond is a long-term effort oriented to the development to coherent sources of "hard" radiation, which may have a number of applications in imaging and spectroscopy. Frequency upshift is also a necessary way to shorten the pulse duration down in the attosecond regime and to obtain higher intensities, as a high frequency pulse might be focused in an extremely small volume. The ultimate frontier of the race towards higher and higher intensities may be considered the Schwinger field $E_s = m_e^2 c^3 / e\hbar$ that has been already introduced at the end of Sect. 2.1.5.

Nonlinear optical effects in plasmas may be exploited to manipulate intense laser pulses. In a broad sense the use of plasma as an optically active and controllable medium is unavoidable at high intensity, since any material is ionized instantaneously by the laser field already at intensities two orders of magnitude lower than the relativistic threshold $a_0 = 1$. Exploiting nonlinear optical properties in the relativistic regime $a_0 \gg 1$ has been named as "relativistic engineering". The moving mirror, already described in Sect. 5.7.1, offers a paradigm for extreme spectral and intensity modulation. For instance, if a laser pulse impinges on a counterpropagating mirror with $\beta \to 1$, then both the frequency and the intensity of the reflected pulses are increased by a factor $(1+|\beta|)/(1-|\beta|) \simeq 4\gamma^2 \gg 1$. Since the number of cycles in the pulse is a Lorentz invariant, the reflected pulse is also shortened by the same factor.

A. Macchi, *A Superintense Laser-Plasma Interaction Theory Primer*,
SpringerBriefs in Physics, DOI: 10.1007/978-94-007-6125-4_6,
© The Author(s) 2013

In the next two Sections we describe two schemes for frequency upshift and amplification of intense laser pulses which are both based on the moving mirror concept. First we describe the *flying mirror* concept which is based, in a broad sense, on the generation of thin and relativistically fast plasma sheets for back-reflection of an external source pulse. We focus on the realization of flying mirrors in an underdense plasma, exploiting the reflection and focusing from nonlinear wake waves (Sects. 3.6.1, 3.6.2). Then we discuss *high harmonic generation* in an overdense plasma, where the reflecting surface performs oscillations driven by the Lorentz force of the laser pulse (see Sect. 4.2) resulting in the modification of the spectral and temporal structure of the reflected pulse.

6.2 Flying Mirrors

In Sect. 4.1 we have mentioned that the laser wake wave in a plasma has phase velocity v_p close to c and, moreover, the phase fronts of the density perturbation have a parabolic shape. Since a parabola should focus perfectly incoming radiation, the nonlinear wake wave is inspiring as a moving mirror or a *flying mirror*, quoting Bulanov et al. (2003) who proposed this concept (sketched in Fig. 6.1) for pulse amplification.

Let us consider a highly nonlinear, close to breaking wave produced by a driver laser pulse of frequency ω_d in an underdense plasma with $\omega_p \ll \omega_d$. The phase velocity is $v_p \simeq (1 - \omega_p^2/2\omega_d^2)$ corresponding to the relativistic factor $\gamma_p \simeq \omega_d/\omega_p \gg 1$. A second laser pulse of frequency ω_s is injected in the direction counterpropagating to the wake wave, and is partially reflected from the first density front acting as a moving mirror, being at the same time compressed in time by a factor $\simeq 4\gamma_p^2$. The combined effect of pulse compression and of field amplification lead to an enhancement of the reflected intensity by a factor $(1+\beta_p)^2/(1-\beta_p)^2 \simeq (4\gamma_p^2)^2$. In addition, the pulse is focused by the parabolic density shell. A parabola is in principle able to focus close to the diffraction limit, thus if D is the effective diameter of the portion of a light beam which is focused by the parabola (with D being limited by the radius either of the beam or of the parabola), focusing leads to an intensity increase of $\sim(D/\lambda')^2$ where λ', is the wavelength in the frame where the parabola is at rest,

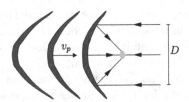

Fig. 6.1 Flying mirror concept. A nonlinear wake wave with parabolic phase fronts reflects a counterpropagating laser pulse, focusing it in space and time

i.e. the wake frame in our case. Thus, the achievable spot size decreases by a factor $\lambda/\lambda' = (1 + \beta_p)\gamma_p \simeq 2\gamma_p$ in the wake frame, and also in the laboratory frame because lengths in the transverse direction are not changed in a Lorentz boost transformation. Hence, the Doppler effect and the focusing by the parabola allows for an overall increase of the intensity by a factor $\simeq (4\gamma_p^2)^2(2\gamma_p)^2(D/\lambda)^2 = 64\gamma_p^6(D/\lambda)^2$.

This very large factor is partly compensated by other effects, the most important one[1] being probably the small reflectivity of the flying mirror. To estimate the latter we consider the parabola as a very thin plasma foil and use the model of Sect. 3.4.2. Following Bulanov et al. (2003), we consider the pulse amplitude not to be relativistic in the wake frame, so that for the reflectivity we have

$$R \simeq \left| \frac{\zeta}{1+\zeta} \right|^2, \quad \zeta = \pi \frac{(n_f d)}{n'_c \lambda'} = \frac{\omega_p^2 d}{2\omega'_s c}, \tag{6.1}$$

where $n_f d$ is the surface density of the foil (a Lorentz invariant) and $\omega' \simeq 2\gamma_p \omega_s$ is the wave frequency in the frame where the mirror is at rest, i.e. in the wake frame. For the surface density, assuming the wake to be close to breaking, we estimate $n_f d \simeq (n_0/2)\lambda_p$ with $\lambda_p \simeq 4(c/\omega_p)\sqrt{2\gamma_p}$ (see Sect. 3.6.2) Thus

$$\zeta \simeq \frac{1}{2} \frac{(\omega_p^2/2)\lambda_p}{2\gamma_p \omega_s c} = \frac{1}{\sqrt{2\gamma_p}} \frac{\omega_p}{\omega_s} = \frac{1}{\sqrt{2\gamma_p^3}} \frac{\omega_d}{\omega_s}, \tag{6.2}$$

where we used $\gamma_p = \omega_d/\omega_p$. We thus obtain for the reflectivity

$$R \simeq |\zeta|^2 = \frac{1}{2\gamma_p^3} \left(\frac{\omega_d}{\omega_s} \right)^2. \tag{6.3}$$

The overall reflectivity *in the laboratory frame*, defined as the ratio between the intensity at the focus I_f and that of the incoming beam I_s, is obtained multiplying R by the enhancement factor due to pulse focusing and compression, obtaining

$$R_{\text{lab}} = \frac{I_f}{I_s} \simeq R \times 64\gamma_p^6 \left(\frac{D}{\lambda} \right)^2 = 32\gamma_p^3 \left(\frac{\omega_d}{\omega_s} \right)^2 \left(\frac{D}{\lambda} \right)^2. \tag{6.4}$$

Assuming "conservative" values such as $\gamma_p \simeq 10$, $D/\lambda \simeq 10$, and $\omega_d/\omega_s \simeq 1$ we get $R \simeq 10^{-3}$ and $R_{\text{lab}} \simeq 2 \times 10^6$. This would increase the intensity of a $I_s = 10^{17}$ Wcm^{-2}, $\lambda_s = 1\,\mu$m pulse (for consistent values of ζ) up to 2×10^{23} W cm^{-2}, beyond present-day limits. Bulanov et al. (2003) consider more challenging parameters such as $D \simeq 200\lambda$ and $\gamma_p \simeq 100$ to achieve 0.8×10^{29} W cm^{-2}, which would bring us close to the Schwinger field.

---

[1] A second effect, not considered here but mentioned by in Bulanov et al. (2003), is the "relativistic aberration" due to the fact that the rays of light reflecting at an angle will have different frequencies.

However, reaching such impressive fields might be hard not only because of technical reasons (i.e. building up a laser system large enough to provide a plasma wave of 200 μm diameter, which would imply a considerable amount of energy) but also because of other detrimental effects at play. Issues not considered in the original paper are related to nonlinear effects in the interaction of the colliding pulse with the mirror: the source pulse might either slow down the mirror by the effect of radiation pressure, or have its transmission coefficient increased by relativistic transparency; this latter effect might limit the pulse amplitude to $a_0 \simeq \zeta$ (notice that a_0 is invariant), and this could spoil the γ_p^3 coefficient in the amplification factor. Another issue of possible relevance in high frequency shift conditions is that coherent reflection from the mirror requires that $n_e' \lambda'^3 > 1$ in the wave frame, and this condition might be problematic when $\gamma_p \sim 10^2$. Additional detrimental effects, such as thermal broadening of the density spike (lowering its reflection coefficient) and small-scale transverse inhomogeneities (reducing the reflected pulse focusability) are discussed in Solodov et al. (2006).

Although approaching QED fields might remain too challenging, the flying mirror scheme could be an interesting approach for producing ultrashort XUV pulses. Experimentally, so far a frequency upshift by a factor in the $37-66$ range (corresponding to $\gamma_p = 3-4$) and a mirror reflectivity $R = (0.3-2) \times 10^{-5}$ have been demonstrated at a counterpropagating pulse intensity of about $I_s \simeq 10^{17}\,\mathrm{W\,cm^{-2}}$ (Kando et al. 2009), in fair agreement with the theoretical expectations.

Several studies have proposed to generate flying mirrors for back-reflection of a second pulse using thin solid foils. With respect to the underdense case, the higher density leads to a larger reflection coefficient, but both driving the mirror to highly relativistic speeds and bending it for focusing is more challenging. In these studies the flying mirror is provided either by the whole foil, accelerated by the driver pulse as in the light sail concept (Sect. 5.7.3) (Esirkepov et al. 2009; Ji et al. 2010) or only by the electrons which are detached by the ions and form a thin electron sheet (Kulagin et al. 2007; Wu et al. 2010). Recently, a scheme where an intense attosecond pulse is generated using a relativistic electron sheet without the need of a second source pulse has been also described (Wu and Meyer-ter-Vehn 2012). In the near-field region such pulse appears as an asymmetric, "half cycle" oscillation of the EM field.

6.3 High Harmonic Generation

In Sect. 5.7.1 we have reviewed the moving mirror concept to describe the acceleration at the surface of an overdense plasma, driven by the radiation pressure of a laser pulse. In such case we have focused on the steady motion of the surface, which is appropriate for circular polarization of the pulse. For linear polarization, however, as discussed in Sect. 4.2 the Lorentz force of the laser pulse drives longitudinal oscillations at the surface, either at the laser frequency ω or at 2ω depending on the angle of incidence and whether the polarization is S or P. Thus, we expect the plasma surface to behave as an *oscillating mirror*. Such model accounts for *high harmonics*

Fig. 6.2 The oscillating mirror model. Figure adapted from Tsakiris et al. (2006)

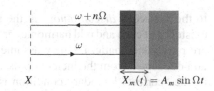

(HH) generation in the spectrum of the reflected light (Bulanov et al. 1994; Linde and Rzàzewski 1996; Lichters et al. 1996; Tsakiris et al. 2006).

We now show that when a wave of frequency ω impinges on a mirror oscillating at frequency Ω, as depicted in Fig. 6.2, the reflected wave will contain multiple frequencies $\omega + n\Omega$, with $n = 1, 2, \ldots$. Let us refer to Fig. 6.2 and to normal incidence for simplicity. The mirror motion is described by the function $X_m(t)$. Let us consider an incident "photon" (i.e. a phase front) which at the time t crosses the plane $x = X$, where we measure an incident field $E_i \sim \sin \omega t$. Such photon will reach the mirror and be reflected at a time t' given by $t' = t + (X + X_m(t'))/c$, and will be back at the $x = X$ plane at $t'' = t + 2(X + X_m(t'))/c$, where we now measure a reflected field $E_r \sim \sin(\omega(t + 2(X + X_m(t'))/c)$, apart from a phase factor. Now, let us assume that the mirror oscillates with $X_m = A_m \sin(\Omega t)$, and for the moment that $A_m \ll \lambda$, so that we may further assume that $X_m(t') \simeq X_m(t)$ (retardation effects will be discussed below). Thus, we can write for the temporal dependence of E_r in X

$$\sin\left(\omega t + \frac{2\omega}{c} A_m \sin \Omega t\right) = \mathrm{Im}\left(e^{-i\omega t} \sum_{n=-\infty}^{n=+\infty} J_n\left(\frac{2\omega}{c} A_m\right) e^{-in\Omega t}\right)$$

$$\sim \sum_{n=0}^{n=+\infty} J_n\left(\frac{2\omega}{c} A_m\right) \sin((\omega + n\Omega)t), \qquad (6.5)$$

where we used the Jacobi identity $e^{-iz \sin \alpha} = \sum_{n=-\infty}^{n=+\infty} J_n(z) e^{-in\alpha}$ and the $J_n's$ are Bessel functions.

If the mirror is driven by the incident pulse itself, its dominant oscillation frequency will be $\Omega = \omega$ in the case of oblique incidence and P-polarization; in such case, we obtain a reflected spectrum containing *all* the HH of the incident frequency, i.e. $\omega, 2\omega, \ldots$. In the other cases, the mirror is driven at $\Omega = 2\omega$, thus we obtain a frequency combination $(2n + 1)\omega$ that means that only *odd* HH will be generated.

The same conclusion about the spectrum of the reflected wave may be obtained by imposing a boundary condition at the moving surface. If the surface is perfectly reflecting in its rest frame, it is easy to show that the condition that the transverse electric field $E'_\perp(X'_m) = 0$ in such frame implies that in the laboratory frame $A_\perp(x = X_m(t), t) = 0$, where $A_\perp$ is the vector potential. Writing $A_\perp = A_{i,\perp} + A_{r,\perp}$ with $A_{i,\perp}$ the incident wave (e.g. $A_{i,\perp} = A_0 \exp(ikx - i\omega t)$) one obtains again (6.5). Notice that often the boundary condition has been written as $E_\perp(x = X_m(t), t) = 0$, that is not correct for a perfect moving mirror for which the electric field component parallel to the surface is *discontinuous* at the surface. However, the conclusions about the existence of even and odd harmonics are not affected. Moreover, the above arguments are purely kinematic, i.e. they assume the motion of the mirror to be independent and decoupled from the incident pulse, which is not the case of interest. The choice of the "correct" boundary condition is not sufficient to obtain a more quantitative information on the reflected spectrum, that requires a self-consistent model for the mirror motion driven by the incident pulse as discussed below.

Additional features of the reflected spectrum and of the resulting waveform may be estimated on the basis of the moving mirror formulas $\omega_r = \omega(1+\beta)/(1-\beta)$ and $E_r = E_i(1+\beta)/(1-\beta)$ (see Sect. 5.7.1). Since the mirror velocity $\beta_m = V_m/c$ is synchronized with the oscillations of the incident pulse, the latter will be modulated periodically in intensity and duration, as if the sinusoidal shape was alternatively stretched and compressed. The reflected pulse will then acquire a spiky structure. Moreover, roughly assuming that a given harmonic is generated at some instant due to the corresponding instantaneous velocity of the surface, we expect the HH spectrum to extend up to the cut-off frequency $\omega_{co} \simeq \omega(1+|\beta_{max}|)/(1-|\beta_{max}|)$, with β_{max} the peak oscillation velocity. A similar conclusion is obtained by taking into account retardation effects, that is due when the motion becomes relativistic ($|\beta_{max}| \to 1$). The generalized phase of the reflected pulse may be written as $\varphi_r = \omega t + (2\omega/c)X_m(t')$ as above, except that now the mirror coordinate depends on the retarded time $t' = t - X_m(t')$. An "instantaneous" frequency may be defined as $d_t\varphi_r$. Using $dt/dt' = 1 - \beta_m(t')$ with $c\beta_m = d_t X_m$, we obtain

$$\frac{d\varphi_r}{dt} = \omega + 2\omega\frac{dX_m}{dt} = \omega + 2\omega\beta_m(t')\frac{dt'}{dt} = \omega\frac{1+\beta_m(t')}{1+\beta_m(t')} \le \omega\frac{1+\beta_{max}}{1-\beta_{max}}. \qquad (6.6)$$

For $|\beta_{max}| \to 1$, we obtain $\omega_{co} \simeq 4\omega\gamma_{max}^2$. This may be converted in a rough scaling with the laser amplitude a_0 by assuming $\beta_{max} \simeq a_0/(1+a_0^2/2)^{1/2}$, following considerations similar to those at the end of Sect. 4.2.2.

It may be interesting to test such somewhat heuristic predictions with a toy model that includes the mirror dynamics self-consistently and may be easily implemented and studied numerically. For simplicity we refer to normal incidence. Following the analysis made in Sects. (4.2.3) and (5.7.1), we generalize the equation for a moving mirror by inserting an harmonic recoil force characterized by a frequency ω_p, so that the plasma response is included phenomenologically:

$$\frac{d}{dt}(\gamma_m\beta_m) = \frac{2I_0}{\sigma c^2}(1+\cos(2\omega t'))\frac{1-\beta_m}{1+\beta_m} - \omega_p^2 X_m/c, \qquad \frac{dX_m}{dt} = \beta_m c, \quad (6.7)$$

where $t' = t - X_m/c$. Figure 6.3 shows results of a numerical solution of (6.7), showing the position $X_m(t)$ and the corresponding spectrum of the reflected light.

The synchronization between the incident pulse and the mirror motion implies that the HH are coherent and phase-locked, i.e. the differences between phase factors of different harmonics are not random. Thus, roughly speaking, it is possible from the HH spectra to obtain pulses of very short duration, i.e. a few attoseconds or

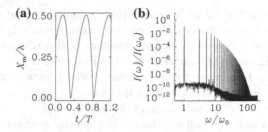

Fig. 6.3 Numerical solution of the "toy" model for HH generation at the surface on an overdense plasma, Eq. (6.7). Frame **a** shows the surface position $X_m(t)$ and **b** the corresponding HH spectrum. Courtesy of S. De Camillis

even below. In fact, it was early realized that the reflected radiation, in appropriate conditions, appears as a train of spikes of attosecond duration (Plaja et al. 1998). To isolate a single attosecond pulses, several approaches has been proposed. These latter include spectral filtering to select the desired harmonics (see e.g. Tsakiris et al. 2006), "polarization gating" by the use of a second pulse to control the polarization of the driving field (Baeva et al. 2006), and manipulation of the wave front of the incident pulse to separate attosecond burst in space (Vincenti and Quéré 2012). Details may be found in reviews on HH generation (Teubner and Gibbon 2009; Thaury and Quéré 2010) and on the more general field of attosecond pulse generation (Krausz and Ivanov 2009; Sansone et al. 2011). The coherence of HH also implies that they may be spatially focused. It has been suggested that due to the favorable scaling of conversion efficiency with the laser intensity, HH focusing in space and time may provide another way to reach the Schwinger limit (Gordienko et al. 2005). It has been argued that the focusability may be reduced by those effects already mentioned for the flying mirror in underdense plasmas (Solodov et al. 2006); however, first experimental results appear to be promising (Nomura, et al. 2009; Dromey et al. 2009).

More complex analytical modeling and PIC simulations of HH generation may be found in the literature. For example, in the relativistic regime "universal" spectra in which the intensity of the nth harmonic in the spectrum falls as n^{-q} have been theoretically predicted with either $q = 5/2$ (Gordienko et al. 2004) or $q = 8/3$ (Baeva et al. 2006). Experimental observations have been in fair agreement with these predictions (Dromey et al. 2006), although the universality of such findings has been questioned (see Boyd and Ondarza-Rovira 2008 and related discussions). In Baeva et al. (2006) it is also found that at ultra-high intensities the cut-off frequency $\omega_{co} \sim \gamma_{max}^3$, a faster scaling that predicted by the oscillating mirror model. This result is reminiscent of the cut-off frequency for synchrotron radiation emission, where the γ^3 scaling is related to the effective duration of the emission bursts in a narrow cone (Jackson 1998, 14.4), and in fact the ultrarelativistic HH regime has been named as *coherent synchrotron emission* (CSE) (van der Brügge and Pukhov 2010). The origin of CSE is deeply related to the oscillations of highly relativistic electron bunches at the plasma surface, which gives rise to radiation emission also in forward direction (i.e. transmitted through the target) as was recently experimentally observed (Dromey et al. 2012).

A connection between the oscillating mirror dynamics and the nonlinear oscillations leading to electron acceleration near the surface (Sect. 4.2) is apparent, and in fact the pulsed acceleration of electron bunches is also responsible for a third mechanism for HH, *coherent wake emission* (CWE) (Quéré et al. 2006). A detailed theory of CWE has been presented in a recent tutorial review (Thaury and Quéré 2010). Briefly, the electrons accelerated by the "vacuum heating" mechanism excite plasma wakes (Sect. 4.1.1) as they re-enter the overdense plasma. The electrostatic plasma wakes may turn into radiating modes in a density gradient, as explained in Thaury and Quéré (2010), and coherent HH are generated. It turns out that CWE is most efficient at moderate intensities, $a_0 \lesssim 1$.

References

Baeva, T., Gordienko, S., Pukhov, A.: Phys. Rev. E **74**, 065401, 046404 (2006)
Boyd, T.J.M., Ondarza-Rovira, R.: Phys. Rev. Lett. **101**, 125004 (2008)
Bulanov, S.V., Naumova, N.M., Pegoraro, F.: Phys. Plasmas **1**, 745 (1994)
Bulanov, S.V., Esirkepov, T., Tajima, T.: Phys. Rev. Lett. **91**, 085001 (2003)
Dromey, B., et al.: Nat. Phys. **2**, 456 (2006)
Dromey, B. et al., Nat. Phys. **5**, 146 (2009)
Dromey, B., et al.: Nat. Phys. **8**, 804 (2012)
Esirkepov, T.Z., Bulanov, S.V., Kando, M., Pirozhkov, A.S., Zhidkov, A.G.: Phys. Rev. Lett. **103**, 025002 (2009)
Gordienko, S., Pukhov, A., Shorokhov, O., Baeva, T.: Phys. Rev. Lett. **93**, 115002 (2004)
Gordienko, S., Pukhov, A., Shorokhov, O., Baeva, T.: Phys. Rev. Lett. **94**, 103903 (2005)
Jackson, J.D.: Classical Electrodynamics, 3rd edn. Wiley, New York (1998)
Ji, L.L., et al.: Phys. Rev. Lett. **105**, 025001 (2010)
Kando, M., et al.: Phys. Rev. Lett. **103**, 235003 (2009)
Krausz, F., Ivanov, M.: Rev. Mod. Phys. **81**, 163 (2009)
Kulagin, V.V., Cherepenin, V.A., Hur, M.S., Suk, H.: Phys. Rev. Lett. **99**, 124801 (2007)
Lichters, R., Meyer-ter-Vehn, J., Pukhov, A.: Phys. Plasmas **3**, 3425 (1996)
Nomura, Y., et al., Nat. Phys. **5**, 124 (2009)
Plaja, L., Roso, L., Rzazewski, K., Lewenstein, M.: J. Opt. Soc. Am. B **15**, 1904 (1998)
Quéré, F., et al.: Phys. Rev. Lett. **96**, 125004 (2006)
Sansone, G., Poletto, L., Nisoli, M.: Nat. Photon. **5**, 655 (2011)
Solodov, A.A., Malkin, V.M., Fisch, N.J.: Phys. Plasmas **13**, 093102 (2006)
Teubner, U., Gibbon, P.: Rev. Mod. Phys. **81**, 445 (2009)
Thaury, C., Quéré, F.: J. Phys. B: At. Mol. Opt. Phys. **43**, 213001 (2010)
Tsakiris, G.D., Eidmann, K., Meyer-ter-Vehn, J., Krausz, F.: New J. Phys. **8**, 19 (2006)
van der Brügge, D., Pukhov, A.: Phys. Plasmas **17**, 033110 (2010)
Vincenti, H., Quéré, F.: Phys. Rev. Lett. **108**, 113904 (2012)
von der Linde, D., Rzàzewski, K.: Appl. Phys. B: Laser and Optics **63**, 499 (1996)
Wu, H.C., Meyer-ter-Vehn, J., Fernández, J., Hegelich, B.M.: Phys. Rev. Lett. **104**, 234801 (2010)
Wu, H.C., Meyer-ter-Vehn, J.: Nat. Photon. **6**, 304 (2012)